Underst

MW01534006

OSHA ELECTRICAL DESIGN SAFETY STANDARDS

Including the Official Regulations

Intertec Electrical Group
New York, NY   Overland Park, KS

Published by
EC&M
INTERTEC Electrical Group
888 Seventh Avenue
New York, NY 10106

ISBN 0-87288-458-9

Please Note: The designations "National Electrical
Code", "NE Code", and "NEC", where used in this
book, refer to the National Electrical Code®, which is
a registered trademark of the National Fire Protection
Association.

CONTENTS

PAGE

Preface vi

Chapter 1
INTRODUCTION 1
OSHA performance standards — an analogy 1
History 3
OSHA experience with the NE Code 5
Part I regulations — what they cover 7
Part I rules — organization 9

Chapter 2
SEC. 1910.302 —
ELECTRIC UTILIZATION SYSTEMS 11
Sec. 1910.302(a) — Scope 11
Sec. 1910.302(b) — Extent of application 15

Chapter 3
SEC. 1910.303 —
GENERAL REQUIREMENTS 17
Sec. 1910.303(a) — Approval 17
Sec. 1910.303(b) — Examining equipment 17
Sec. 1910.303(c) — Splices 19
Sec. 1910.303(d) — Arcing parts 19
Sec. 1910.303(e) — Marking 19
Sec. 1910.303(f) — Identification of disconnect
                              means 20
Sec. 1910.303(g) — Clearances 20
Sec. 1910.303(h) — Over 600V 23

Chapter 4
SEC. 1910.304—
WIRING DESIGN AND PROTECTION 29

Sec. 1910.304(a) — Grounded and grounding
conductors 29
Sec. 1910.304(b) — Branch circuits 29
Sec. 1910.304(c) — Open conductors (600V or less)....30
Sec. 1910.304(d) — Services 34
Sec. 1910.304(e) — Overcurrent protection................36
Sec. 1910.304(f) — Grounding 37

Chapter 5
SEC. 1910.305 —
WIRING METHODS, COMPONENTS, AND
EQUIPMENT 45

Sec. 1910.305(a) — Wiring methods 45
Sec. 1910.305(b) — Cabinets, boxes, and fittings........49
Sec. 1910.305(c) — Switches 50
Sec. 1910.305(d) — Switchboards and panelboards 50
Sec. 1910.305(e) — Enclosures for damp or wet
locations 51
Sec. 1910.305(f) — Conductors for general wiring 51
Sec. 1910.305(g) — Flexible cords and cables 51
Sec. 1910.305(h) — Portable cords over 600V 53
Sec. 1910.305(i) — Fixture wires................................53
Sec. 1910.305(j) — Equipment for general use 53

Chapter 6
SEC. 1910.306 —
SPECIFIC PURPOSE EQUIPMENT AND
INSTALLATIONS 61

Sec. 1910.306(a) — Electric signs and outline lighting .61
Sec. 1910.306(b) — Cranes and hoists 62

Sec. 1910.306(c) — Elevators, escalators, and
moving walks 64
Sec. 1910.306(d) — Electrical welders 64
Sec. 1910.306(e) — Data-processing systems 66
Sec. 1910.306(f) — X-ray equipment 66
Sec. 1910.306(g) — Induction and dielectric heating ..67
Sec. 1910.306(h) — Electrolytic cells...........................68
Sec. 1910.306(i) — Electrically driven irrigation
machines 71
Sec. 1910.306(j) — Swimming pools, fountains, etc. ..71

Chapter 7
SEC. 1910.307 —
HAZARDOUS (CLASSIFIED) LOCATIONS 75
Sec. 1910.307(a) — Scope 75
Sec. 1910.307(b) — Electrical installations 81
Sec. 1910.307(c) — Conduits 84
Sec. 1910.307(d) — Equipment in Div. 2 locations 84

Chapter 8
SEC. 1910.308 —
SPECIAL SYSTEMS...87
Sec. 1910.308(a) — Systems over 600V nominal.........87
Sec. 1910.308(b) — Emergency power systems 90
Sec. 1910.308(c) — Class 1, 2, and 3 circuits 91
Sec. 1910.308(d) — Fire protective signaling systems ..93
Sec. 1910.308(e) — Communications systems.............94

Chapter 9
SEC. 1910.309 — NATIONALLY RECOGNIZED
TESTING LABORATORIES 99

Appendix
OSHA SUBPART S — Part I
OFFICIAL REGULATIONS 105

PREFACE

S

INCE THE FIRST EDITION of EC&M's "Illustrated Guide to OSHA Electrical Rules" was published in 1982, the Occupational Safety and Health Administration (OSHA) has recast its rules and set out to draw up four sets of electrical safety regulations that directly affect electrical systems designers, engineers, contractors, installers, and workers.

This book covers Part I of these OSHA regulations: "Design Safety Standards for Electrical Installations" which became law in April, 1981. A future EC&M book will explain Part II, "Electrical Safety-Related Work Practices". The other two, Part III, "Safety-Related Maintenance Requirements," and Part IV, "Safety Requirements for Special Equipment," will follow when OSHA publishes these rules.

Parts I-IV are contained in Subpart S, Part 1910 of Title 29 of the Code of Federal Regulations (CFR).

In addition, OSHA has devised electrical safety regulations specifically for construction sites. A future EC&M book will also explain Subpart K, "Electrical Standards for Construction". Subpart K is in 29 CFR Part 1926.

OSHA regulations, like those of every other U.S. government department, are officially published in the Federal Register with differing effective dates. The OSHA books offer rule-by-rule synopses and explanations of the rules from the register.

A complete copy of the regulations as they appear in the Federal Register is included in an Appendix for quick reference. To make the text easier to correlate with the interpretations, the sections and subsection designations in the actual rules have been highlighted by the use of bold type to match those in Chapters 1 through 8. Those are the only changes that have been made to the text of the actual rules.

New York, NY
April, 1991

Brendan B. Read
Assistant Editor,
Electrical Code Watch
and EC&M

The problem of protecting workers from electric shock has existed from the beginning of domestic and commercial electrical applications. Employees for the early electric street railway utility companies and electrical contractors faced a strange new danger when attaching "hot" overhead wire to support brackets, fixing generators, and installing and repairing fuse panels in homes and offices. The National Electrical Code (NEC) was created in the 1896 out of concerns for electrical safety.

The best way to think of the OSHA Part I regulations is that they are the "NEC of safety in the workplace." These rules have been drawn up and enforced by the federal Occupational Safety and Health Administration (OSHA), which has the responsibility to help ensure a safe work environment. The agency sets performance standards, but unlike the NE Code, its rules do not directly tell an employer how to meet them. As long as items and installations meet a given criteria, it is up to the employer to decide how the rules will be met. This is different from the implementation of the NE Code, where inspectors from the "authority having jurisdiction" follow the stipulations in the code to see if the work has been done to the section specifics.

OSHA performance standards — an analogy

A regulatory comparison can be made with highway speed limits in the U.S. Let's say you are anxious to get to a location. The weather is clear and the road is dry. A state trooper then pulls you over and tickets you for going 65 mph in a 55 mph zone. Regardless of the road conditions that could have let you zip up to 75 mph safely, you broke the highway **code.**

Now envision the same situation with the weather rainy, a strong wind blowing, and the roadway slick. In order to be on the jobsite on time you drive at your normal 55 mph and are stopped by the trooper. Why? Because you were violating the **performance standard** of the road. Or as the officer puts it,"driving without due care to conditions." You should have been doing 35 instead of 55.

That is how an OSHA inspector will look at electrical equipment once it's installed and up and running. Does it meet the design and safety criteria at a workplace? This book, which is part of *EC&M's* series on OSHA electrical rules, will help you comply.

It is up to the employer to think safety first, last, and always and to continually use common sense. The OSHA regulations in this book will not tell you to how to provide a safe workplace or serve as a stipulation-by-stipulation safety code. These regulations will tell you what OSHA and American workers have a right to expect in the way of electrical safety.

A little variation in electrical safety performance can have serious and sometimes tragic results. Health and safety experts say that electric shock currents as low as 3 milliamperes can cause serious indirect injuries, such as bone fractures and even death, by involuntary muscle action. The National Center for Health Statistics reports that approximately 1,000 people are accidentally electrocuted every year. Serious injuries such as electric burns caused by current flowing in the tissue take a long time to heal. Electrical workers are also exposed to arc and thermal burns caused by high-temperature arcs and flashes, such as when switchgear shorts out or when they can come into contact with the hot surfaces of overheating electrical equipment. Clearly, effective and evolving safety performance standards that are followed by employers and employees are needed to reduce the amount of resulting human tragedy and cost.

History

When the federal Occupational Safety and Health Act that created OSHA became law in 1970, the new agency had to come up with standards to make the workplace safer. It created Part 1910, "General Industry Standards." The agency then devised Subpart S to cover electrical safety.

Part I of Subpart S outlines electrical safety rules for installations. These OSHA regulations cover only those electrical system parts that employees use or come into contact with. The rules require that these systems be made and installed in a manner that will minimize risk to people. The reasoning is simple: if electrical installations are put in with safety in mind to begin with, there will be less injury and death.

OSHA adopted the only readily available electrical design safety standard—the NEC. The original OSHA Part 1910, Subpart S, Part I promulgated in 1972, referenced the 1971 NE Code and included many definitions used in the code. In a departure from federal government administrative procedures, and out of concern for workplace safety, Congress authorized OSHA to adopt and

Our understanding of safety has changed over time as new techniques have been developed for protecting persons against accidental contact with energized parts of an electrical system. The open-front switchboards have given way to metalclad switchgear, which has greatly improved safety.

apply many NE Code provisions retroactively. The code itself is not a retroactive document. However, had OSHA not done this, and chose to apply the 1971 Code only to new installations, employees at existing workplaces would be unprotected.

By basing its regulations on the NE Code, OSHA drew on the code's extensive experience. It is one of the most respected set of safety rules in the world and is designed to be readily enforced by electrical inspectors. The NE Code consensus code-making process incorporates many years of expertise from all sectors of the electrical industry: electricians, contractors, inspectors, utilities, and manufacturers.

While the NE Code gave OSHA a supportable set of workplace installation standards with Part I, OSHA's imprimatur expanded and solidified the code's status as the one recognized American body of electrical rules. Federal government recognition and enforcement of the NE Code in the workplace prompted many local jurisdictions to adopt the code.

OSHA experience with the NE Code

OSHA learned through experience that the NE Code alone is not flexible enough to meet its enforcement and regulatory needs. Its standards needed to accommodate changes in technology without constant rewriting, and to allow alternative installation techniques if they provide the same safety level as the code. This way OSHA can maintain and enhance its safety performance criteria with a minimum of paperwork and problems for employers and employees. Referencing the NE Code, which changes every three years, would force the federal agency to modify its regulations on the same cycle. OSHA says that such an approach is not practical because of the length of time it takes to do so.

While the NEC is an excellent design document, it does not contain comprehensive enough standards for safety-related work practices, electrical maintenance, or special equipment. The code was originally written to protect against fire caused by electricity, not necessarily to protect employees. In 1976 the National Fire Protection Association (NFPA), which sponsors the NEC, formed the NFPA 70E committee, "Electrical Safety Requirements for Employee Workplaces" to draft a consensus standard that would help OSHA prepare regulations. The 70E committee broke down each of these regulations into Parts I-IV. OSHA Subpart K, which sets electrical standards for construction, is derived from Part I.

This book on Part I, *Electrical Design Standards*, is drawn from the 70E committee's work on the 1978 NEC. The 70E standards, which have the 1978 NE Code as its source, greatly simplified the essence of the code for OSHA's purposes by deleting topics and tables that had little to do with occupational safety. OSHA keeps requirements such as conductors having to be protected with approved insulation, but stays away from the specifics. That the equipment operates safely is OSHA's main concern.

At the same time, the 70E committee strengthened employee safety requirements by adding provisions covering electrically driven and/or controlled irrigation equipment, electrolytic cell use at workplaces, and fire protective signaling systems. There are possible shock hazards to workers with all three of these types of electrical technology and installations.

The NFPA makes the distinction between the NE Code and OSHA regulations very clear. In the foreward to the 70E standard it points out that:

"The NEC is intended for use primarily by those who design, install, and inspect electrical installations. OSHA's electrical regulations address the employer and employee in their workplace. The technical content and complexity of the NEC is extremely difficult for the average employer and employee to understand.

"Some of the detailed provisions within the NEC are not directly related to employee safety, and therefore are of little value for OSHA's needs."

OSHA published a revised Part 1910, Subpart S, Part I, *Design Safety Standards for Electrical Systems*, based on the 70E Part I standards. It went into effect in April, 1981. The document had been substantially changed from the original 1971 NEC-based OSHA Part I. The revised regulations contained only 15,000 words compared to 250,000 words in the NE Code. Its scope follows the code in that it covers the electrical installation from the service drop or the utility connection to the load and back, including the ground. Part I also freed employers and OSHA from having to reference the NE Code for workplace safety. This way, Part I will remain current as the code and NFPA 70E change.

Part I regulations — what they cover

The 1981 Part I rule has had two major changes made to it since it came into effect.
• Construction site rules that were covered in Sec. 1910.304(b)(1) in the 1981 rule are now in part of a separate document, Subpart K.
• References to Underwriters Laboratories (UL) and Factory Mutual (FM) in the Sec. 1910.399(a) "Definitions" as being examples of "nationally-recognized testing laboratories" (NRTLs) have been deleted. Instead, OSHA refers to another document, Part 1910 Sec. 1910.7. Chapter

Modern production facilities such as this one for producing paper contain many potential sources of accidents due to the great number of motors and other electrical equipment involved. Following the OSHA regulations on design safety standards for electric utilization systems will greatly reduce the risks.

9 in this book covers NRTLs and what they test for.

These regulations, like other OSHA regulations, can be strengthened or modified by individual states. Thusfar, 25 states and territories have done so. While 19 of these have programs generally identical to OSHA's, 6 states (Alaska, California, Hawaii, Michigan, Oregon, and Washington) have regulations that are frequently not identical to, but as or more effective than the OSHA rules.

For example, Article 26, Sec. 2940 of California's *Elec-*

trical Safety Orders require a qualified or in-training employee to act as an observer to work being performed on exposed high-voltage conductors and equipment. This rule serves to help prevent accidents and/or give assistance to an injured worker.

The 19 states and territories whose rules are generally identical to OSHA's are: Connecticut, Indiana, Iowa, Kentucky, Maryland, Minnesota, Nevada, New Mexico, New York, North Carolina, Puerto Rico, South Carolina, Tennessee, Utah, Vermont, Virginia, Virgin Islands, and Wyoming. Note that Connecticut and New York's rules cover only state and local government employees.

Part I rules — organization

OSHA has organized the entire Part 1910, Subpart S to accomodate not only Part I but also Parts II, III, and IV so as to provide a systematic format for the agency's entire planned set of electrical design safety rules. It has kept certain sections open to facilitate any new rules that may be added later.

The organization is as follows.

Sec. 1910	Subject
.301	Introduction
.302	Electric utilization systems
.303	General requirements
.304	Wiring design and protection
.305	Wiring methods, components, and equipment for general use
.306	Specific purpose equipment and installations
.307	Hazardous (classified) locations

.308   Special systems
.309-.330   Design safety standards (reserved)
.331-.360   Safety-related work practices
.361-.380   Safety-related maintenance requirements
.381-.398   Safety requirements for special equipment
.399   Definitions
Appendix A   Reference documents
Appendix B   Expanatory data (reserved)
Appendix C   Tables, notes, and charts (reserved)

This section defines the scope of the OSHA Part I regulations, which is very similar to that in the NE Code. It also sets out which rules are retroactive.

Retroactivity means that employers must modify, if necessary, the electrical equipment or installation at the workplace to comply with the rules regardless of when the installation was made or the equipment placed. OSHA sets out two lists: one which shows the rules that are retroactive, and the other which shows the rules that apply to electrical systems put in after April 16, 1981, the day this Part I regulation came into force. The rules that are retroactive are interspersed throughout the entire set of regulations.

The intent of retroactivity is to cover employees working on all electrical installations. In contrast with the OSHA rules, NE Code Sec. 90-1(b) does not distinguish between existing and new installations. When Part I was being drafted, a number of respondents noted that the NE Code was not meant by its authors to apply retroactively. If retroactivity was meant to be in the Code, the writers could have refined NE Code Sec. 90-2 "Scope" to include that provision.

Sec. 1910.302(a) — Scope

The facilities covered by Secs. 1910.302 through 1910.308 includes electrical systems used in buildings, structures, and premises including yards, carnivals, parking and other lots, mobile homes and RVs, industrial

Food distribution centers are a far cry from the warehouses of old. The electrical systems within these facilities and virtually all other have become much more extensive and complex. OSHA regulations provide design guidelines to improve the safety of such systems.

substations, other outside premises wiring, and conductors that connect them to the power supply.

Not covered by these regulations are:

• electrical installations on aircraft, motor vehicles, trains, ships, and other transportation means (except mobile homes and RVs);

• installations in underground mines;

• generation, distribution, signaling, and communication installations of railroads;

• communication equipment installations under the

Secs. 1910.302 through 1910.308 of the OSHA electrical safety requirements do not cover installations under the exclusive control of electrical utilities for the generation, transmission, and distribution of electrical power.

exclusive control of communication utilities, whether located outdoors or in building spaces used exclusively for the purpose;

• installations under the exclusive control of electrical utilities for generation, transformation, transmission and distribution, communications, or metering, whether located in buildings used exclusively by utilities for such purposes, or located on property owned or leased by the utility, or on public highways, streets, etc., or on established rights-of-ways on private property.

Sec. 1910	Subject
.303(b)	Examination, installation, and use of equipment
.303(c)	Splices
.303(d)	Arcing parts
.303(e)	Marking
.303(f)	Identification of disconnecting means
.303(g)(2)	Guarding live parts
.304(e)(1)(i)	Protection of conductors and equipment
.304(e)(1)(iv)	Location in or on premises
.304(e)(1)(v)	Arcing or suddenly-moving parts
.304(f)(1)(ii)	2-wire DC systems to be grounded
.304(f)(1)(iii)	AC systems to be grounded
.304(f)(1)(iv)	" " " " "
.304(f)(1)(v)	AC systems 50 to 1000V not required to be grounded
.304(f)(3)	Grounding connections
.304(f)(4)	Grounding path
.304(f)(5)(iv)(a) through	Fixed equipment required to be grounded
.304(f)(5)(iv)(d)	" " " " " "
.304(f)(5)(v)	Grounding of equipment connected by cord and plug
.304(f)(5)(vi)	Grounding of nonelectrical equipment
.304(f)(6)(i)	Methods of grounding fixed equipment
.305(g)(1)(i)	Uses of flexible cords and cables
.305(g)(1)(ii)	" " " " " "
.305(g)(1)(iii)	Flexible cords and cables prohibited
.305(g)(2)(ii)	Flexible cord and cable splices
.305(g)(2)(iii)	Pull at joints and terminals of flexible cords and cables

Table 2.1. Retroactive regulations that apply to all installations regardless of when they were installed.

Sec. 1910-302(b) — Extent of application

Part I rules have been modified over the length of time in which they have been in force. Thus, there are certain rules that apply to all installation, and others that apply retroactively only to installations built after specific dates. OSHA has listed these in tables for easy reference.

Requirements contained in the sections listed in **Table 2.1** are applicable to all electrical installations regardless of the date when they were designed or installed.

All electrical utilization systems and pieces of equipment that were installed after March 15, 1972 must comply with all the provisions of Secs. 1910.302 through 1910.308, not just the provisions of those items listed in **Table 2.1**. In addition, all major replacement, modification, repair, or rehabilitation done after March 15, 1972 on the systems and equipment installed before March 15, 1972 are also required to meet all the provisions of Secs. 1910.302 to 1910.308. Major replacements, modifications, or rehabilitation is defined in the OSHA regulations as including work similar to that involved when a new building or facility is built, a new wing is added, or an entire floor is renovated.

A major revision of the OSHA Part I regulations became effective as of April 16, 1981. There are certain parts of revised Secs. 1910.302 through 1910.308 that are intended only to apply to those systems and equipment installed after April 16, 1981. These non-retroactive sections are listed in **Table 2.2**.

Sec. 1910	Subject
.303(h)(i)	Entrance and access to workspace over 600V
.303(h)(ii)	" " " " " " "
.304(e)(1)(vi)(b)	Circuit breakers operated vertically
.304(e)(1)(vi)(c)	Circuit breakers used as switches
.304(f)(7)(ii)	Grounding of systems of 1000V or more supplying portable or mobile equipment
.305(j)(6)(ii)(b)	Switching series capacitors over 600V
.306(c)(2)	Warning signs for elevators and escalators
.306(i)	Electrically controlled irrigation machines
.306(j)(5)	Ground-fault circuit interrupters for fountains
.308(a)(1)(ii)	Physical protection of conductors over 600V
.308(c)(2)	Marking of Class 2 and Class 3 power supplies
.308(d)	Fire-protective signaling circuits

Table 2.2. Regulations that only apply to those electrical installations and equipment that was installed after April 16, 1981.

This section sets out overall requirements for electrical systems. It covers installation and equipment approval, examination, installation and use. It details stipulations for splicing, marking, disconnecting means and circuit identification, working clearances, clear spaces, illumination, live parts guarding, and installation enclosures.

Sec. 1910.303(a) — Approval

Conductors and equipment that are required or permitted in Subpart S (Electrical) of the OSHA regulations are acceptable only if approved. In Sec. 1910.399, "Definitions," the meaning of "approved" is given as "acceptable to the authority enforcing this Subpart...the Assistant Secretary of Labor for Occupational Safety and Health." Thus, conductors and equipment are acceptable if they are either accepted, certified, listed, labeled, or found to be safe by an OSHA-accredited Nationally-Recognized Testing Laboratory (NRTL) (see Chapter 9). OSHA must approve labs that test workplace products to recognized safety standards.

Sec. 1910.303(b) — Examining equipment

Electrical equipment must be free from hazards that could cause injury and death. Safety or danger of a particular piece of equipment can be determined by looking for the following.
• Suitability as evidenced by its being listed or labeled.

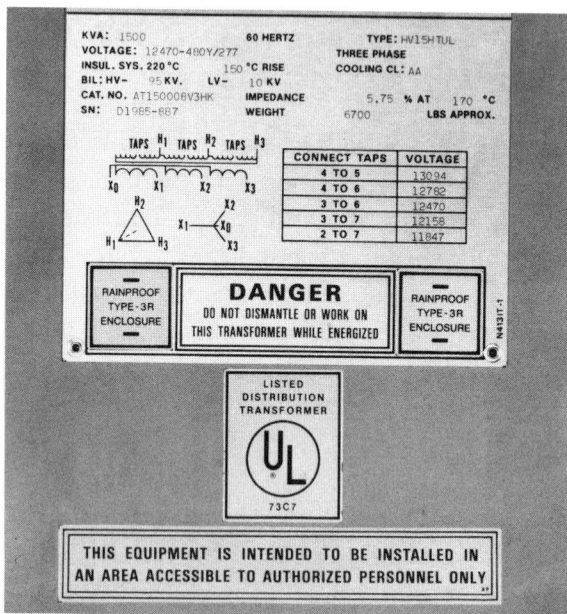

KVA: 1500 — 60 HERTZ — TYPE: HV15HTUL
VOLTAGE: 12470-480Y/277 — THREE PHASE
INSUL. SYS. 220 °C — 150 °C RISE — COOLING CL: AA
BIL: HV— 95 KV. LV— 10 KV
CAT. NO. AT150008V3HK — IMPEDANCE — 5.75 % AT 170 °C
SN: D1985-887 — WEIGHT — 6700 — LBS APPROX.

TAPS H1 TAPS H2 TAPS H3

CONNECT TAPS	VOLTAGE
4 TO 5	13094
4 TO 6	12782
3 TO 6	12470
3 TO 7	12158
2 TO 7	11847

RAINPROOF TYPE-3R ENCLOSURE

DANGER
DO NOT DISMANTLE OR WORK ON
THIS TRANSFORMER WHILE ENERGIZED

RAINPROOF TYPE-3R ENCLOSURE

LISTED DISTRIBUTION TRANSFORMER

UL 73C7

THIS EQUIPMENT IS INTENDED TO BE INSTALLED IN
AN AREA ACCESSIBLE TO AUTHORIZED PERSONNEL ONLY

One of the ways allowed by the OSHA rules for determining whether a piece of electrical equipment is free from hazards is by looking for evidence that it has been listed or labeled.

- Strength and durability of the equipment enclosure.
- Condition of electrical insulation.
- Overheating during use.
- Signs of arcing.
- Type, size, voltage, and current-carrying capacity match application needs.
- Other factors needed in the workplace to safeguard employees.

Listed or labeled equipment must be installed in accordance with the instuctions included in the listing or labeling.

Sec. 1910.303(c) — Splices

Conductor splicing must be accomplished with devices suitable for the purpose, or by brazing, soldering, or welding. Soldered splices have to be mechanically and electrically secure before solder is applied. Insulation or insulating devices that provide the same protection as that on conductors must cover all joints and splices.

Sec. 1910.303(d) — Arcing parts

Separation and isolation from all combustible material is required for electrical equipment that usually arcs, flames, sparks, or produces molten metal. One way of accomplishing this is by enclosing the equipment in an enclosure suitable for the purpose.

Sec. 1910.303(e) — Marking

The marking rule requires that electrical equipment used in the workplace have voltage, current, wattage, and other necessary ratings noted on the equipment. The marking itself must be able to withstand workplace use so that its information is available to employees. The markings can either be placed on at the factory or at the installation when equipment is custom-built at a jobsite.

Note that OSHA's meaning of "marking" is different from that of "labeling" in the NE Code. Equipment that is labeled means that the unit complies with the appropriate electrical standards according to a particular testing laboratory. The marking required by OSHA is for ready identification of the suitability of the equipment for the application and atmosphere encountered in the workplace. The manufacturer's name, trademark or other identification marking is also required to be placed on the equipment.

Sec. 1910.303(f) — Identification of disconnect means

Every required disconnect means of motors and appliances must be legibly marked to indicate its purpose. Each service, feeder, and branch circuit must be similarly marked. The only exception is if the purpose of the circuit is clearly evident. The marking must be sufficiently durable for the environment in which the disconnect is located.

Sec. 1910.303(g) — Clearances

This set of regulations applies to installations that operate at 600V nominal or less, and covers working clearances and live parts guarding. Sufficient access and working space must be provided and maintained around all electrical equipment.

Fig. 3.1 shows the dimension of the working clearances required by OSHA in front of live parts that require adjusting, servicing or maintenance. The measurement is taken from the live part if exposed, or from the enclosure front or opening. In (A), exposed live parts are on one side and no live or grounded parts on the other side of the workspace. Concrete, brick, or tile walls are considered to be grounded. This spacing also applies if live parts exist on both sides of the workspace and are separated by wood

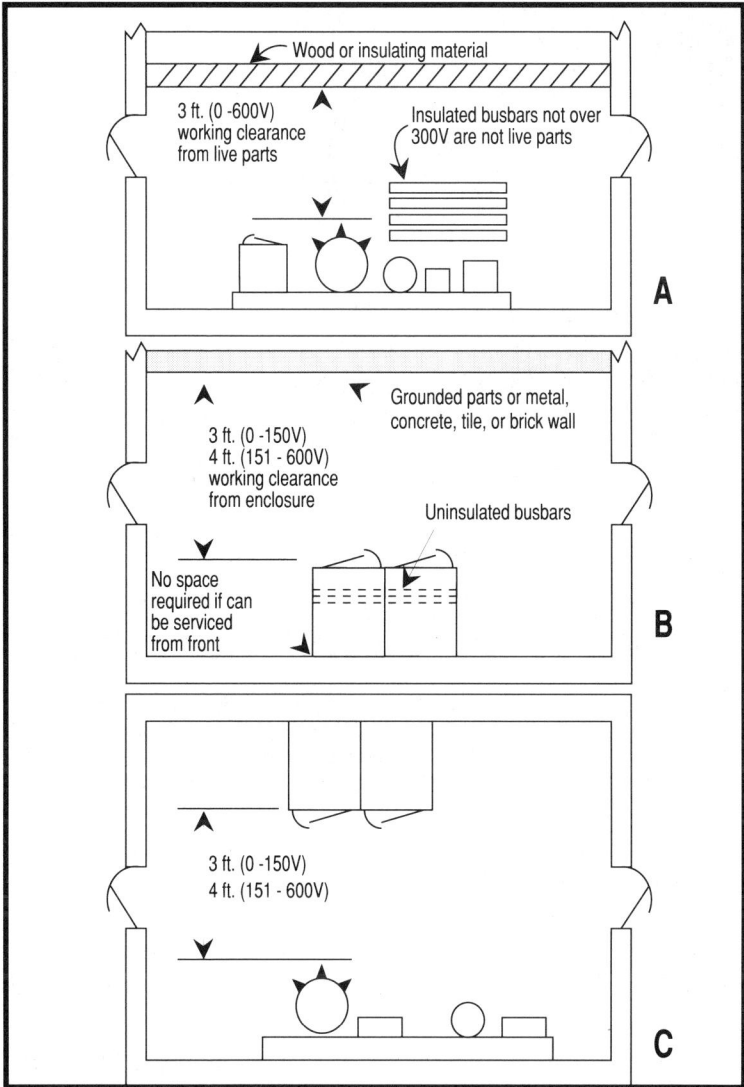

Fig. 3.1. Working space requirements around electrical equipment operating in 0-600V range depend upon various factors.

Disconnect means are required to be identified so that in an emergency a person can readily determine what each controls and can take quick action if required to deenergize the circuit.

---

or other insulated material. The working clearance required if there are live parts on one side and grounded surfaces on the other is shown in (B). Note that for enclosed live parts the clearance is measured from the enclosure front, not from the live parts. In (C), unguarded live parts exist on both sides of the workspace.

No clearance is required in back of electrical equipment

where all the renewable or adjustable parts, and connections are accessible from the front or other locations. A minimum headroom of 6ft-3in. must be maintained within the working spaces. For installations built before April 16, 1981, the minimum working clearance requirement is 2ft-6in.

The working space requirements are not permitted to be encroached upon. No storage is to take place within the working space. Suitable barriers must be established around the working clearance space when live parts of normally-enclosed equipment is exposed. This protects against close approach by persons unfamiliar with the dangers of electrical equipment. Lighting must also be provided for the working space around equipment installed indoors.

Live parts of equipment operating at 50V or more must be guarded to prevent accidental contact by:

- using a suitable enclosure;
- locating within a room or vault accessible only to qualified persons, and marked with a conspicuous warning sign forbidding unqualified persons from entering;
- surrounding with partitions or screens that prevent touching with the body or with a metal probe;
- locating on a balcony or elevated platform arranged to exclude unqualified persons;
- or locating at an elevation of 8 ft or higher.

Sec. 1910.303(h) — Over 600V

Provisions of Secs. 1910.303(a) through (g) also apply to systems operating at over 600V. The following set of regulations points out the additional rules that apply to these systems. They, however, do not apply to equipment on the supply side of the service conductors. A medium-voltage installation is potentially more hazardous when compared with low-voltage installations. Therefore, access to such equipment is permitted to only qualified persons. To be considered accessible to qualified persons only, the installation must be in a vault, room, closet, or an area surrounded by a wall, screen, or fence, and access must be controlled by lock and key or other approved means. A wall, screen, or fence less that 8-ft tall is not considered as capable of preventing access unless other means are employed to make it equivalent. The enclosure doors must be kept locked or under the observation of a qualified person.

When the equipment is within an area that is accessible to unqualified persons, the live parts must be metal-enclosed or located in a locked vault. The metal-enclosed equipment must be kept locked if the bottom is less than 8 ft above the floor. All must be provided with caution signs, must be protected against physical damage, and prevent foreign objects inserted through ventilation openings from contacting live parts.

Workspace must be allowed about the equipment as indicated in **Fig. 3.2** and be wide enough to allow the equipment doors or hinged panels to swing open to 90°. In (A), exposed live parts are on one side and no live or grounded parts on the other (or live parts on both sides effectively guarded by insulating material). The working clearance required if there are live parts on one side and grounded surfaces (including concrete, brick, or tile walls) on the other is shown in (B). In (C), unguarded live parts exist on both sides of the workspace.

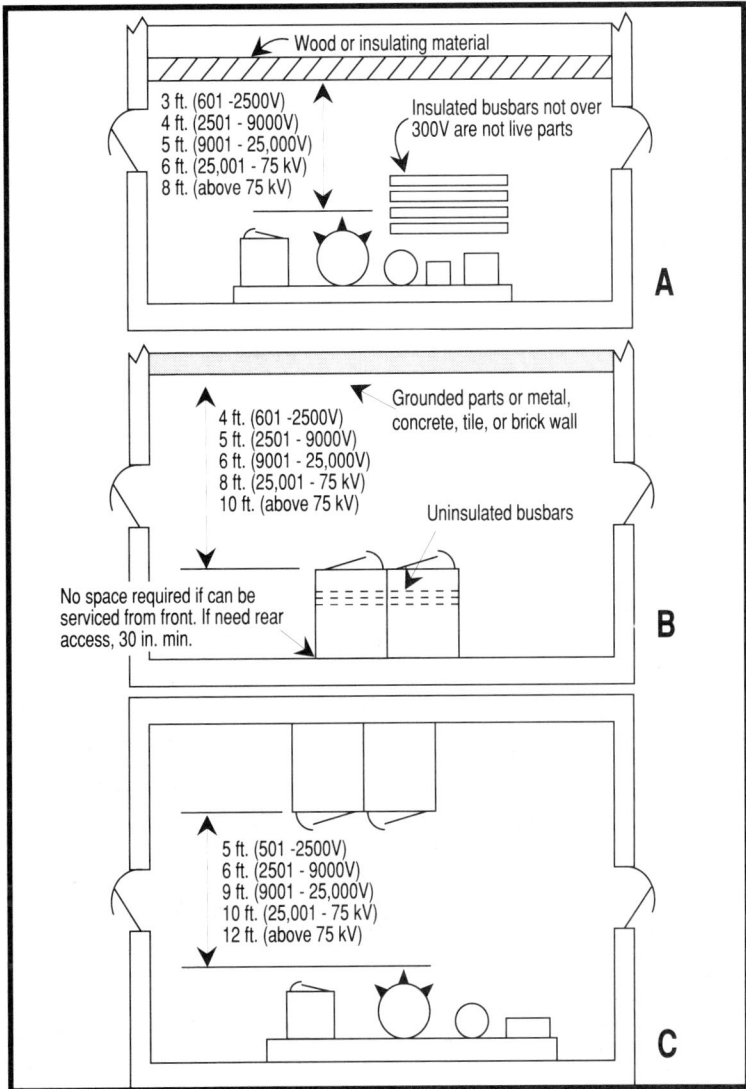

Wood or insulating material

3 ft. (601 -2500V)
4 ft. (2501 - 9000V)
5 ft. (9001 - 25,000V)
6 ft. (25,001 - 75 kV)
8 ft. (above 75 kV)

Insulated busbars not over
300V are not live parts

A

4 ft. (601 -2500V)
5 ft. (2501 - 9000V)
6 ft. (9001 - 25,000V)
8 ft. (25,001 - 75 kV)
10 ft. (above 75 kV)

Grounded parts or metal,
concrete, tile, or brick wall

Uninsulated busbars

No space required if can be
serviced from front. If need rear
access, 30 in. min.

B

5 ft. (501 -2500V)
6 ft. (2501 - 9000V)
9 ft. (9001 - 25,000V)
10 ft. (25,001 - 75 kV)
12 ft. (above 75 kV)

C

Fig. 3.2. Working space requirements around electrical equipment operating at over 600V varies according to the voltage level and other factors.

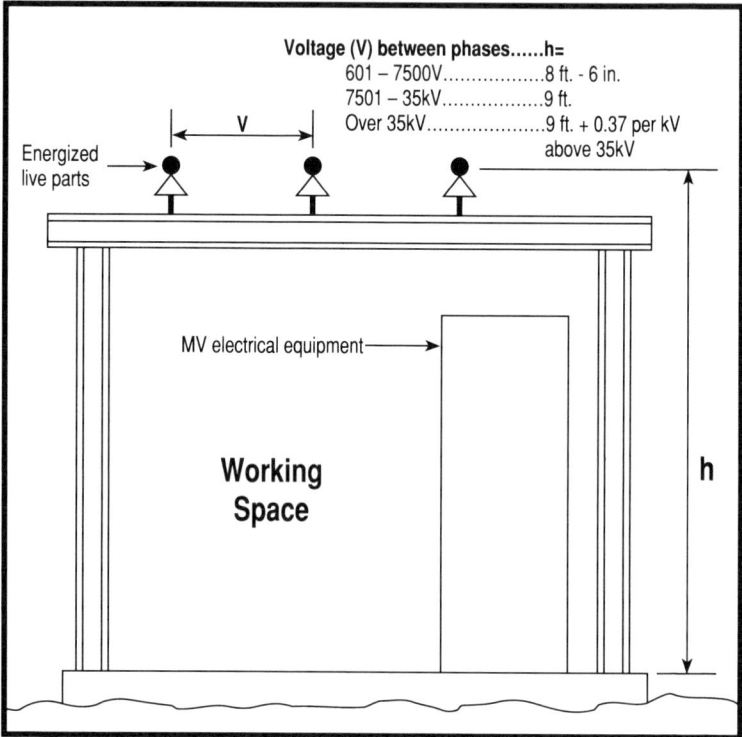

Fig. 3.3. Required minimum elevation of unguarded energized parts (over 600V) above a working space depends upon the nominal voltage level of the live parts.

The minimum workspace height is 6ft-6in. and the area must have adequate illumination. The lighting fixtures and switches must be located so that persons changing the lamps or turning on the lights are not in danger of contacting live parts. For equipment installed prior to April 16, 1981 operating above 25kV, the required working clearance does not have to be increased beyond that shown as required for 25kV systems.

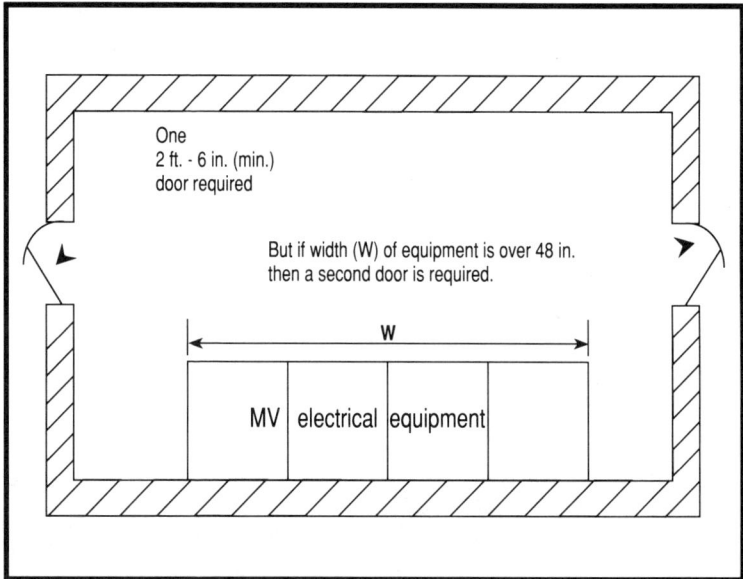

Fig. 3.4. Required entrance and access to workspace. This provision applies to installations built after April 16, 1981.

There are also minimum height clearances required for unguarded energized parts above working spaces. These are shown in **Fig. 3.3**. For installations in place before April 16, 1981, the minimum clearance height is 8 ft, but this is limited to those cases where the voltage between phases is between 601 and 6600V.

As shown in **Fig. 3.4**, the number of access doors required to rooms or vaults containing medium-voltage electrical equipment depends upon the width of the electrical equipment within the room. If the width is over 48 in., then a second door is required. Bare energized live parts that are adjacent to a door must be suitably guarded to prevent accidental contact by persons entering or exiting the area. If the room is on a mezzanine, roof, etc., permanent stairs or ladders must be installed to permit safe access.

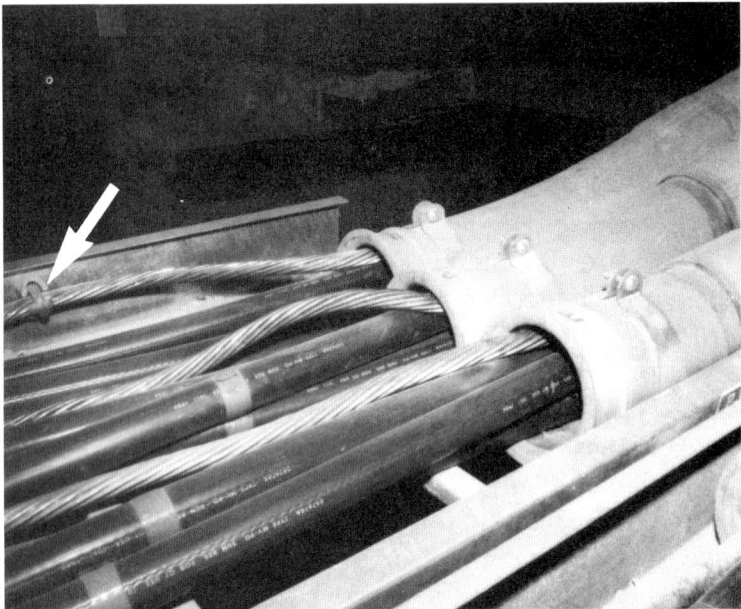

Grounding conductors must be identifiable and distinguishable from all other conductors. This OSHA safety rule helps to ensure that this vital system is not damaged when work is performed on wiring networks.

This section addresses wiring systems from the service conductors to branch circuits. It also lists requirements for conductor overcurrent and physical protection.

Sec. 1910.304(a) — Grounded and grounding conductors

The grounded (neutral) conductors and the (equipment and system) grounding conductors must be readily identifiable and distinguishable from all other conductors. Consistent with OSHA policy, these rules leave it to the workplace manager to decide how to meet this performance criterion. However, the NE Code is the best guide to compliance.

Another requirement is that polarity markings must be observed. It is contrary to OSHA regulations to attach a grounded conductor to a terminal intended for a "hot" wire, reversing the polarity. The screwshell of a lighting fixture should never become the normally-energized leg of the circuit. Likewise, the grounding terminal on a receptacle, plug, or connector is forbidden to be used for any other purpose.

Sec. 1910.304(b) — Branch circuits

This subsection formerly contained regulations on branch circuits and outlet devices at construction sites. These have been deleted and the issue addressed in a separate body of regulations, Subpart K, *Electrical Standards for Construction*. These rules can be found in another book in *EC&M's* series on the OSHA regulations.

The only rule remaining in this subsection says that outlet devices must have an ampere rating not less than the load to be served.

Sec. 1910.304(c) — Open conductors (600V or less)

Conductors run outdoors as open conductors on a pole line are required to be spaced far enough apart to allow an authorized person servicing the line adequate horizon-

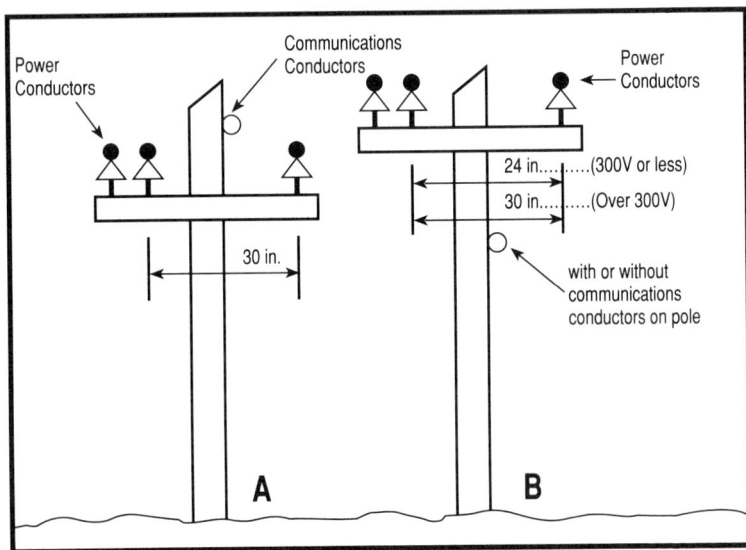

Fig. 4.1. Minimum climbing space requirements for branch circuit, feeder, or service conductors, 600V or less, run outdoors on polelines. In (A), the communications conductors are above the power conductors, and in (B) the power conductors are either above the communications conductors or are alone on the pole.

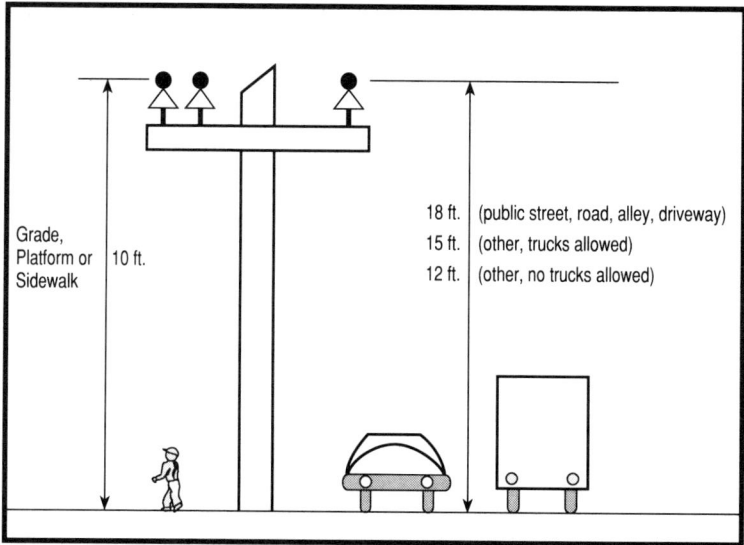

Fig. 4.2. Minimum vertical clearances of overhead open conductors, 600V or less, from grade, sidewalks, platforms, or other projections from which they may be reached.

tal climbing space past the conductors. Open conductors, as used here, is equivalent to the NE Code definition of a "bare" conductor as one having no covering or electrical insulation whatsoever; or of a "covered" conductor as one encased within material of composition and thickness that is not recognized by the code as electrical insulation. **Fig. 4.1** summarizes the minimum requirements.

Vertical clearances of overhead open conductors from grade, sidewalks, platforms from which the conductors could be reached, roads, etc., as shown in **Fig. 4.2,** must be maintained.

When run outdoors as open conductors, power circuits operating at 600V or less must be placed so that they

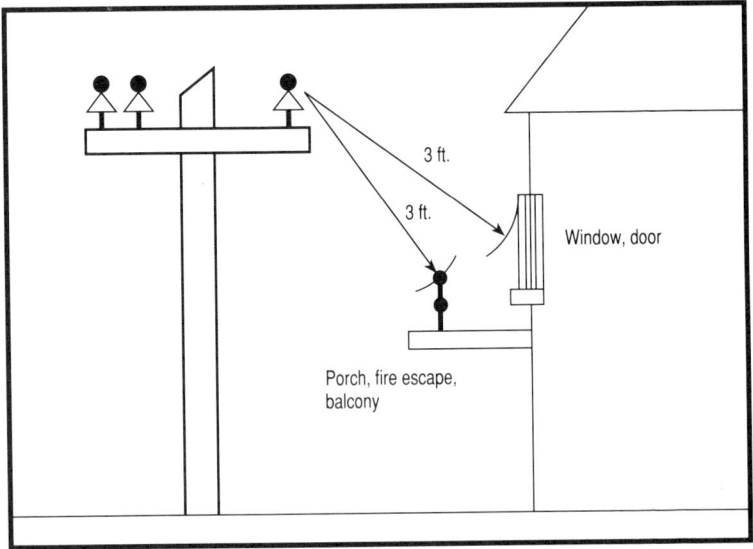

Fig. 4.3. Minimum clearances of overhead open conductors, 600V or less, from building openings such as windows, doors, porches, etc.

cannot be accidentally contacted by persons reaching from a window, balcony, or other building openings. The spacing requirement is shown in **Fig. 4.3.** Conductors run above the top of the window are considered to be out of reach from the window and, therefore, are not subject to the 3-ft clearance rule.

The general rule is that where open conductors pass over a roof, an 8-ft clearance from the highest point on the roof must be allowed (**Fig. 4.4**). However, a different rule applies when the open conductors operating at 300V or less pass over a roof. If the slope of the roof is at least 4-in. in 12, the clearance from the highest point on the roof can be reduced to 3 ft.

If the conductors operating at 300V or less pass over no

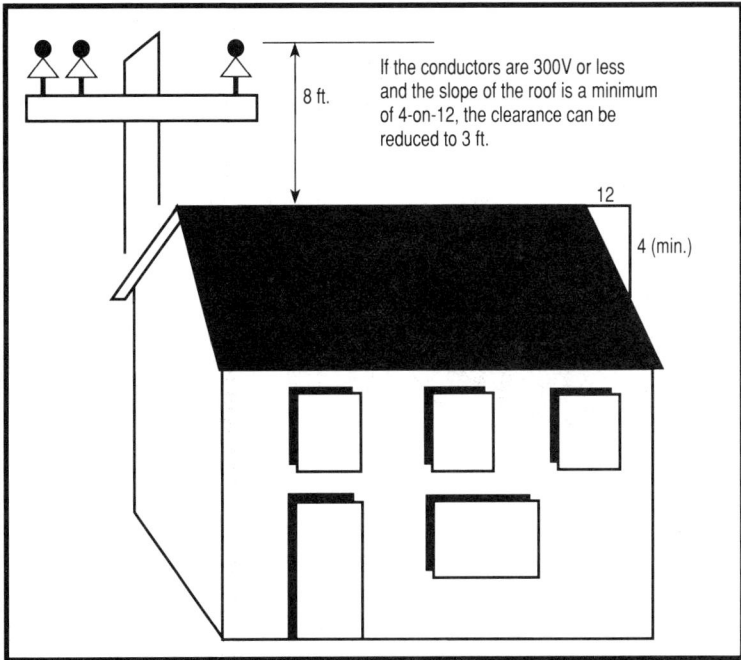

Fig. 4.4. Minimum clearance of overhead open conductors, 600V or less, over roofs.

more than 4-ft of roof overhang and terminate in an approved through-the-roof raceway or support as shown in **Fig. 4.5**, the clearance from the roof must be at least 18 in.

Outdoor lighting fixtures mounted on poles or structures carrying live conductors must be located below the conductors, transformer, capacitors, or other items associated with the power distribution system unless the equipment can be deenergized by a lockable disconnecting means, or if adequate clearances or other safeguards are provided. This provision seeks to assure the safety of

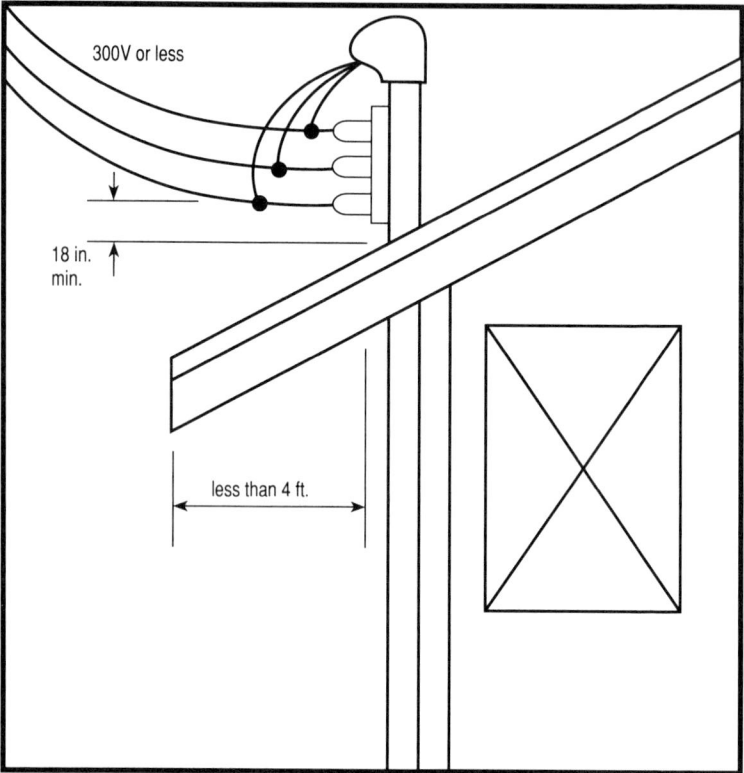

Fig. 4.5. Minimum clearance of conductors, 300V or less, which run over less than 4 ft of a roof overhang and terminate at a through-the-roof raceway or other approved support.

persons relamping the lighting fixture.

Sec. 1910.304(d) — Services

Disconnecting means must be provided to isolate the conductors within a building or structure from the service conductor. The disconnect must be installed nearest the point of entrance of the service conductors, open all ungrounded conductors simultaneously, have its handle positions clearly marked to indicate whether it is open or closed, and it must be located in a readily accessible location. According to the definitions given in Sec. 1910.399, "readily accessible" means that the item is capable of being reached quickly for operation, renewal, or inspections, without requiring those to whom ready

A disconnecting means, located near to the point of entrance of the service conductors, is required to isolate the electrical system within a facility from the service conductors.

access is requisite to climb over or remove obstacles or to resort to portable ladders, chairs, etc.

Services over 600V are also required to meet these rules, and must in addition adhere to several other rules. Service-entrance conductors that consist of open wires must be guarded at the service disconnect to make them accessible only to qualified persons. In addition, warning signs of high voltage must be posted where other-than-qualified persons may come in contact with live parts.

Sec. 1910.304(e) — Overcurrent protection

Conductors and equipment operating at 600V or less must be protected against overcurrent based upon their ability to safely conduct current. NE Code Tables 310-16 through 310-19 are the best guide available for determining safe conductor ampacities.

The overcurrent device must not open the grounded (neutral) conductor unless at the same time it simultaneously opens all the ungrounded conductors. Overcurrent protection of systems operating at over 150V to ground that utilizes fuses is required to include a disconnecting means that will allow the fuses to be deenergized when they are replaced. Opening the disconnect must not interrupt power to unrelated circuits.

Overcurrent devices must be located so that they are readily accessible to each employee or authorized building management personnel. The devices must be kept clear of locations where they are exposed to physical damage, and where they are in the proximity of easily ignitible material. Fuses and circuit breakers must not pose the threat of burning or injuring employees. This can be accomplished by location or shielding.

Circuit breakers must be clearly marked to indicate whether they are open (off) or closed (on). The "on" position must be at the top (unless operation is horizontal or by rotation). If used as switches in 120V fluorescent lighting circuits, the circuit breakers must be approved for the purpose and marked "SWD." These last two provisions apply to equipment installed after April 16, 1981.

Sec. 1910.304(f) — Grounding

The following systems are required to be grounded.

• A 3-wire DC system must have its neutral conductor grounded.

• A 2-wire DC system operating at 50 to 300V between conductors unless: they supply only industrial equipment in a limited area and are supplied with ground detectors; are fire-protective signalling circuits having a maximum current of 0.030A; or are rectifier-derived from AC circuit meeting the requirements for AC systems of 50V or less, and 50-1000V listed below.

• AC circuits less than 50V if supplied by a transformer having an ungrounded primary or a primary voltage over 150V to ground. These circuits must also be grounded if installed as outdoor overhead conductors.

• AC circuits 50 to 1000V when: the system can be grounded and the resulting system will have a maximum voltage to ground of 150V; a 480Y/277V, 3-phase, 4-wire system where the neutral is used as a circuit conductor; a 240/120V, 3-phase, 4-wire system in which the midpoint of one phase is used as a circuit conductor; or if a circuit conductor is insulated.

Grounding is an essential part of electrical safety and is stressed in the OSHA regulations. Noncurrent-carrying metal parts that may become energized, such as this gate of a substation fenced enclosure, must be grounded.

Some exceptions apply to the requirement of grounding AC systems operating at 50 to 1000V. Grounding is not required if: the circuit is used exclusively for industrial electric furnaces; it is separately derived and used only to supply adjustable-speed drives; or it is separately derived from a transformer having a primary voltage rating less than 1000V and the system is used only for control circuits; is part of an isolated power system in a health-care facility; or requires continuity of control power. Ground detectors must be installed and the circuit must be maintained and supervised by qualified persons if not grounded.

For a grounded system, a grounding electrode conductor must be used to connect the equipment grounding conductors and the grounded circuit conductors to the grounding electrode at the supply side of the service disconnecting means, or the supply side of the disconnect means or overcurrent device of a separately-derived system. It is the identified conductor of AC premises wiring systems that must be grounded. This refers to white, natural gray, or white with a colored stripe that identify the conductor as being a grounded (neutral) conductor.

If the system is ungrounded, then the equipment grounding conductors must be connected to the grounding electrode at the service equipment, or ahead of the system disconnecting means or overcurrent device if it is a separately-derived system.

The path to ground from circuits, equipment, and enclosures must be permanent and continuous. If no equipment grounding conductor is present in an existing branch circuit that is to be extended, the grounding terminal of a grounding-type receptacle can be connected to a grounded cold water pipe near the equipment.

Equipment required to be grounded includes the following.

• Metal cable trays, raceways, and enclosures for conductors.

• Service equipment metal enclosures.

• Frames of electric ranges, clothers dryers, and other permanently-connected appliances and the connection boxes that are part of the circuit.

Neither sleeves used to protect cable from physical damage, nor metal enclosures for conductors added to existing open wire, knob-and-tube wiring, and NM cable in runs of 25 ft of less (and which are not likely to contact ground and are guarded against employee contact) need to be grounded.

Exposed noncurrent-carrying metal parts of fixed equipment that may accidentally become energized must also be grounded if it is:

• within 5 ft horizontally or 8 ft vertically of ground or grounded metal objects as shown in **Fig. 4.6**;

• located in a wet or damp location and not isolated;

• in electrical contact with metal;

• in a hazardous (classified) location;

• or is supplied by a metalclad, metal sheathed, or grounded metal raceway wiring method.

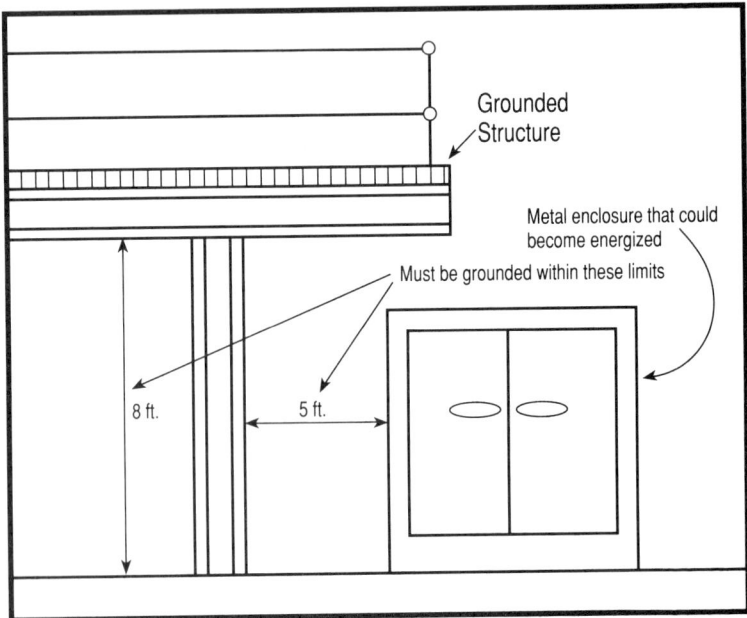

Fig. 4.6. Exposed noncurrent-carrying metal parts of fixed equipment that may become energized is required to be grounded if within these limits from ground or grounded metal objects.

However, if the equipment operates with any terminal at over 150V to ground, then certain classes of equipment do not have to have their metal enclosures grounded. Included are: switches and circuit breakers used for other than service equipment (if accessible to qualified persons only); electrically-heated appliances that are permanently and effectively insulated from ground; and distribution transformers, capacitors, and similar equipment mounted on wooden poles at a height exceeding 8 ft above ground or grade level.

Cord-and-plug connected equipment must have provisions for grounding exposed noncurrent-carrying metal parts that could become energized. The following types of equipment are covered by this requirement.

• Items used in a hazardous (classified) location.

• Items operated at over 150V to ground (except guarded motors and frames of electrically-heated appliances if permanently and effectively insulated from ground).

• Appliances such as refrigerators, dishwashers, hand-held tools, portable and mobile X-ray and associated equipment, hedge clippers, snow blowers, portable hand lamps, tools likely to be used in wet and conductive locations, and appliances used in damp or wet locations or by employees standing on the ground or on metal floors or working inside metal tanks or boilers.

Listed or labeled portable tools and appliances protected by an approved system of double insulation do not need to be grounded. Neither do tools used in wet and conductive locations that are supplied by an isolation transformer with an ungrounded secondary of not over 50V.

Cranes, elevators, and metal partitions must also be

grounded. Frames and tracks of electrically-operated cranes, frames of nonelectrically-driven elevator cars to which electric conductors are attached, and metal enclosures around equipment over 750V between conductors must be grounded.

Noncurrent-carrying metal parts required to be grounded must be grounded by an equipment grounding conductor run in the same raceway or cable enclosing the circuit conductors (except for DC circuits). Electrical equipment is considered to be effectively grounded if secured to, and in electrical contact with a properly grounded metal rack or structure on which it is supported. For installations made before April 16, 1981 only, electrical equipment secured to and in metallic contact with the structural metal frame of a building is also considered to be grounded. Metal car frames of elevators supported by metal hoisting cables operated by grounded elevator machines are also considered to be effectively grounded.

Systems and circuits operating at over 1000V must meet the grounding requirements listed for those operating at less than 1000V. All noncurrent carrying parts of portable and fixed high-voltage equipment, and their associated fences, enclosures, and supporting structures must be grounded. However, equipment that is guarded by location and isolated from ground, such as pole-mounted distribution apparatus mounted at a height exceeding 8 ft, need not be grounded.

Portable and mobile high-voltage equipment, besides the other requirements, must comply with additional grounding rules.

This equipment must be supplied from a system having its neutral grounded through an impedance. If a delta-connected system is used to supply the equipment, a system neutral must be derived. The equipment must be connected by an equipment grounding conductor to the point at which the system neutral impedance is grounded.

Ground-fault detection and relaying must be provided

New construction and major alterations in a facility involves many workplace situations that expose the employee to potential danger.

to automatically deenergize any high-voltage system component that develops a ground fault. The equipment grounding conductor must be continuously monitored in order to automatically deenergize the high-voltage feeder to the portable equipment upon loss of grounding conductor continuity.

The grounding electrode to which the system neutral impedance is connected must be isolated from and separated by at least 20 ft from any other system or equipment grounding electrode. In addition, there should be no direct connection between the grounding electrodes, such as buried pipe, fences, etc.

The focus of this section is equipment installation; its rules are designed to ensure electrical continuity, reduce the possibility of accidental contact and short-circuits, and protect wiring methods and equipment from physical damage.

Sec. 1910.305(a) Wiring methods

General. This section requires that metal raceways, cable armor, and other metal enclosures for conductors, along with the boxes, fittings, and cabinets that are part of the wiring system, be joined together in a manner that assures electrical continuity. No wiring systems are to be installed in ducts used for carrying flammable vapors, dust, or loose stock. Further, no wiring system of any type may be installed in ducts used for vapor removal or for ventilation of commercial-type cooking equipment. This restriction also applies to any shaft containing only such ducts. It should be noted that the provisions stated above do not apply to conductors that are an integral part of factory-assembled equipment.

Temporary power and lighting at voltages of 600V or less is allowed only for building maintenance, repair, remodeling, or demolition. Temporary systems may also be used for experimental or development work. Decorative lighting for seasonal or carnival purposes installed for no more than 90 days falls within these provisions. Temporary wiring over 600V is permitted only for emergencies, experiments, or tests. The flexible, portable nature of

OSHA allows temporary power only for building construction, maintenance repair, and demolition work. It is important to note that in buildings such as the one in the background, designed as residences and therefore not covered by OSHA, if an occupant decides to have a home business and hires an employee to work there (even part-time) that residence portion used for business becomes a workplace. Any electrical upgrading and maintenance, and every other OSHA regulation designed to protect employee safety therefore applies.

these systems make them inherently less safe than permanent installations. That is why they are no substitute for fixed wiring system.

If not subject to physical damage, temporary wiring may be run as open conductors mounted on insulators that are no more than 10 ft apart. Otherwise, the conductors must be run as multiconductor cord or cable assemblies. Regardless of the configuration, these feeders must originate from an approved power outlet or panelboard.

Branch circuits must also originate from an approved panelboard or power outlet. They cannot be installed directly on floors but may, however, be run as open conductors fastened at ceiling height every 10 ft. Each branch circuit must have a separate equipment grounding conductor, unless it is in a metal raceway, and grounding type receptacles must be used and electrically connected to the grounding conductor. Bare conductors and earth returns are not permitted to be used for temporary wiring purposes. Ungrounded conductors of each temporary circuit must have disconnecting switches or plug connectors to facilitate disconnection. Cables and flexible cords must have protection against physical damage, particularly when passing through doorways or other pinch points. Temporary lighting lamps must either be at least 7 ft from the working surface or have guards.

Cable trays. The following wiring methods are permitted in cable trays:
- mineral-insulated metal-sheathed cable (Type MI);
- armored cable (Type AC);
- metal-clad cable (Type MC);
- power-limited tray cable (Type PLTC);
- nonmetallic-sheathed cable (Type NM or NMC);

Cable tray located in an industrial facility where only qualified persons will maintain the system, are permitted to contain single conductor cable 250kcmil or larger.

- shielded nonmetallic-sheathed cable (Type SNM);
- multiconductor service-entrance cable (Type SE or USE);
- multiconductor underground feeder and branch circuit cable (Type UF);
- other factory-made multiconductor cable approved for tray use;
- and approved conduit and raceways with contained conductors.

Single conductor Type RHH, RHW, MV, USE, or THW cable sized 250kcmil or larger can be used only in industrial applications in ladder, trough, or 4-in. ventilated trays. This permission applies only to those industrial sites where electrical service and maintenance is provided by qualified persons.

If exposed to the sun, cables must be sunlight-resistant. Cable tray systems are not permitted in areas where physical damage is possible, such as in hoistways. If installed in hazardous (classified) locations, cable trays must contain cables specifically listed for use in these areas.

Open wiring is only allowed on systems operating at 600V or lower installed in farms and factories, or for services. Open conductors must be rigidly supported on noncombustible, nonabsorbent insulators not in contact with other objects.

Flexible nonmetallic tubing, in lengths not exceeding 15 ft, may be used to enclose conductors in dry locations where not subject to physical damage. The tubing must be secured by straps every 4 ft-6 in. maximum.

Knob-and-tube wiring methods are to be used when conductors are installed through floors, walls, and cross members.

Sec. 1310.305(b) — Cabinets, boxes, and fittings

Abrasion protection must be provided for all conductors entering boxes, cabinets, or fittings. Any unused openings as well as openings through which conductors enter must be closed. Pull boxes, junction boxes, and fittings must be provided with approved covers. If metallic, they must be grounded. Outlet boxes in completed installations must have covers, faceplates, or fixture canopies. Those covers with holes for flexible cord pendants must have a bushing or smooth, well-rounded opening.

Pull and junction boxes for systems over 600V must meet the same requirements as those for use on lower-voltage systems, and must additionally have suitable covers visibly marked "HIGH VOLTAGE." The covers must be securely fastened in place; underground box covers weighing over 100 pounds meet this securely-fastened provision.

Sec. 1910.305(c) — Switches

Single-throw knife switches must be connected so that the blades are de-energized while in the open position and must be mounted so as to prevent switch closure by gravity. Those switches approved for inverted mounting must have a locking device to ensure they will remain in the open position when set to that position.

Double-throw knife switches are permitted to be mounted either vertically or horizontally; however, a locking device to ensure the switch will remain in the open position must be provided when these switches are vertically mounted.

Faceplates for flush-mounted snap switches mounted in ungrounded metal boxes located near conducting surfaces must be made of nonconducting, noncombustible material.

Sec. 1910.305(d) — Switchboards and panelboards

Switchboards with exposed live parts must be located in permanently dry locations and be accessible by only qualified persons. Panelboards must be of a deadfront construction and be mounted in cabinets, cutout boxes, or enclosures approved for this purpose. Those panelboards not of deadfront construction are permitted only where access is limited to qualified persons. The accident risks with these installations are too great for their operation by untrained, unqualified persons. Their exposed knife switches blades must be de-energized when open.

Sec. 1910.305(e) — Enclosures for damp or wet locations

Weatherproof construction must be used for cabinets, cutout boxes, fittings, boxes, and panelboard enclosures installed in wet locations. In other locations, the enclosures must be installed so as to prevent moisture entrance and accumulation. Switches, circuit breakers, and switchboards installed in wet locations must be housed in weatherproof enclosures.

Sec. 1910.305(f) — Conductors for general wiring

General wiring conductors must have insulation approved for the system's required voltage, operating temperature, and installed application. In addition, the insulation must be color coded to identify grounded, ungrounded, and grounding conductors. Other suitable identification means can be used.

Sec. 1910.305(g) — Flexible cords and cables

Flexible cord and cables must be approved and suitable for the location in which they will be installed. Their application may include pendants; fixture wiring; appliance and portable lamp connections; elevator cables; crane and hoist wiring; stationary equipment connections; noise and vibration transmission prevention; appliances where fasteners and connections are designed for removal for maintenance and repair purposes; and data processing cables in approved systems.

Cords used as connections for portable lamps, appliances, interchangeable stationary equipment, or equipment/appliances requiring removal for maintenance or repair must have an attachment plug for connection to an approved outlet.

Flexible cord and cable use is prohibited as a substitute for fixed wiring in structures. It also is not to be:
- installed through walls, ceilings, and floors;
- placed through doorways, windows, and other openings;
- attached to building surfaces;
- or concealed behind ceilings, floors, and walls.

In showcases and window displays, the regulation permits Type S, SO, SJ, SJO, ST, SJT, STO, SJT, SJTO, and AFS cords unless they are used for chain-supported light fixtures or as supply cords for portable lamps and other display merchandise. The cord type, size, and number of conductors must be durably marked on the surface of the cord. The grounding or equipment-grounding conductor in these cords and cables must be readily identifiable from the rest of the conductors.

Flexible cords must be used in continuous lengths without splices or taps. As an exception, No. 12 AWG or larger hard-service cords may be repaired (spliced) if the splice retains the electrical and physical performance characteristics of the original. Strain relief must be provided on flexible cord connections so that tension will not be directly transmitted to joints or terminal screws.

Sec. 1910.305(h) — Portable cords over 600V

Portable cables over 600V for powering portable or mobile equipment must be made with flexible-stranded No. 8 AWG or larger conductors, have shielding if operating at 2kV or higher, and have grounding conductors. Strain relief locking-type connectors, capable of preventing opening or closing while energized, are also required. Splicing is not permitted unless the splicing material is permanently molded, vulcanized, or of another approved type. Termination enclosure access is limited to qualified personnel and must be marked with a high voltage hazard warning.

Sec. 1910.305(i) — Fixture wires

Fixture wires must be approved for use in their specific voltage, temperature and location application. They are permitted only in lighting fixures, for internal wiring or connection to branch circuit wiring, if enclosed and not subject to twisting or bending. Identification is required for a fixture wire that is used for a grounded conductor. They are not permitted as branch circuit conductors except as Class 1 power-limited circuits.

Sec. 1910.305(j) — Equipment for general use

Lighting fixtures, lampholders, etc., must be designed so that employees will not be normally exposed to live parts. Fixtures installed in wet or damp locations must be approved for the purpose. Rosettes and cleat-type lampholders and receptacles are permitted to have exposed live parts, however, if installed a minimum of 8 ft above the finished floor.

Portable-type handlamps with flexible cords must be equipped with handles of molded composition or other approved material, lampholder guards, and screw-shell type lampholders. If installed in wet or damp locations, lampholders must either be of weatherproof construction or installed to prevent the accumulation of water or moisture.

Receptacles, connectors, and plugs must be so constructed as to prevent the connection of different amperage or voltage-rated items. The only exception is the 20A T-slot receptacle that can accept a 15A attachment plug of the same voltage. Those receptacles installed in wet or damp locations must be approved for the application.

Appliances, except for those with high temperature current-carrying parts requiring exposed construction, cannot have energized parts exposed to employees. In addition, they must be marked with voltage and amperage ratings and be provided with disconnecting means.

Motors and controllers. A disconnecting means must be located in sight of the controller. Per **Fig. 5.1**, "in sight from" means that one piece of equipment is visible and not more than 50 ft from the other. A group of coordinated controllers for a multi-motor continuous process machine mounted next to each other requires only a single disconnecting means located next to the group. For motor branch circuits over 600V, the controller disconnecting means may be located out of sight of the controller if the requirements shown in **Fig. 5.2** are met.

Disconnecting means must operate so that all ungrounded poles open simultaneously to disconnect the

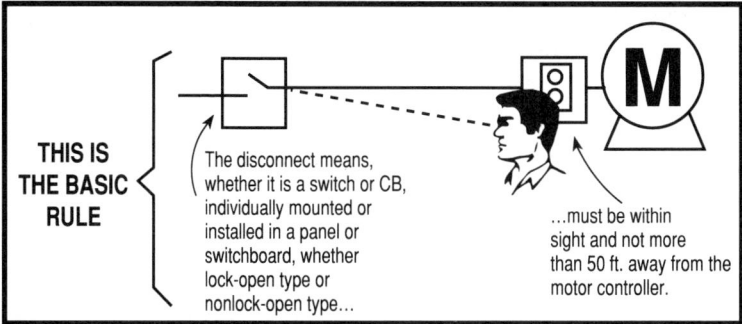

THIS IS THE BASIC RULE

The disconnect means, whether it is a switch or CB, individually mounted or installed in a panel or switchboard, whether lock-open type or nonlock-open type...

...must be within sight and not more than 50 ft. away from the motor controller.

Fig. 5.1. The basic rule for the location of a motor's disconnect means.

motor and its controller, must plainly indicate whether it is in the open or closed position, and must be capable of being locked in the open position.

For those motors not in sight from their controller location, one of the conditions shown in **Fig. 5.3** must be adhered to. If more than one disconnecting means is provided for the same piece of equipment, such as one at

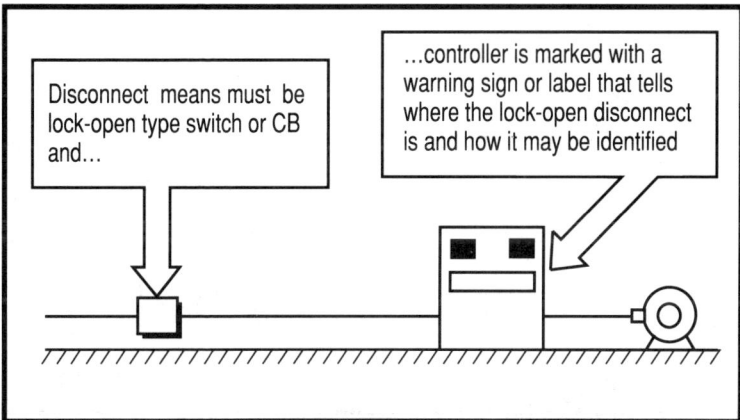

Disconnect means must be lock-open type switch or CB and...

...controller is marked with a warning sign or label that tells where the lock-open disconnect is and how it may be identified

Fig. 5.2. For motors operating at over 600V, the disconnect means can be located out of sight from the controller provided these conditions are met.

the controller and the other at the motor, only one need be readily accessible.

An individual disconnecting means is normally required for each motor. Under the following conditions, however, a single disconnecting means may be used for a group of motors if:

- a number of motors drive special parts of a single machine or piece of equipment as shown in **Fig. 5.4**;
- a group of motors is under the protection of one set of branch-circuit protective device; or
- a group of motors is in a single room in sight from the location of the single disconnecting means.

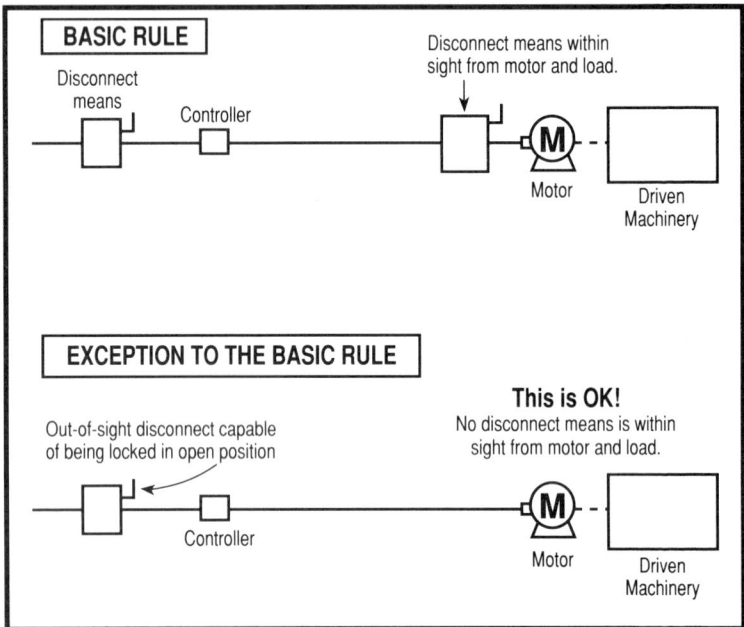

Fig. 5.3. A motor that is out of sight from its controller must meet either one of these conditions.

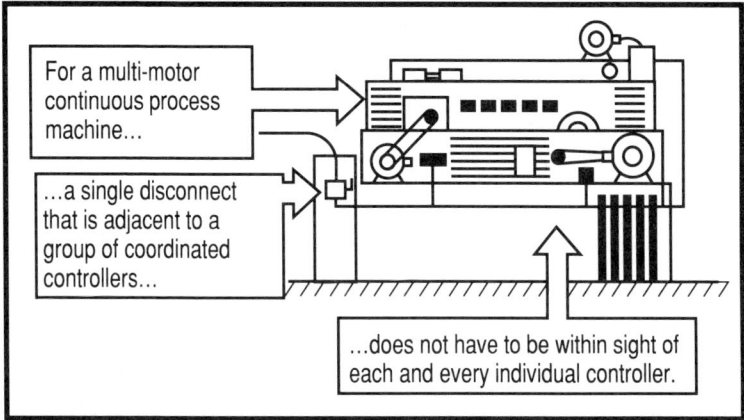

Fig. 5.4. One of the exceptions to the rule that requires an individual disconnect means for each motor.

All motors, motor-control equipment, and motor branch-circuit conductors must be protected against overheating caused by motor overload or failure to start, and short-circuits or ground faults. The only exceptions to these protection rules are: where a shutdown is likely to cause additional hazards, like stopping fire pumps, and where a motor is needed to safely shut down equipment and its overload sensors are connected to a monitored alarm system.

Guarding against accidental employee contact is required for exposed motor and controller parts operating at 50V or greater between terminals. This guarding can be achieved by installing the units in a restricted-access enclosure, room, balcony, gallery, or platform; or by elevating them 8 ft or more off the floor. Exceptions to this guarding requirement are stationary motors with commutators, collectors, and brush rigging located inside the

Liquid-filled transformers are required to be installed in a vault if they present any significant hazard to employees.

---

motor end brackets and not conductively connected to supply power circuits at more than 150V to ground. Insulating mats must be provided at those motors with live motor parts of 150V or more, where their guarding is provided by location only, or where operational adjustments and other personnel may be involved.

Transformers not covered by these regulations are:
- current transformers;
- dry-types that are components of other equipment;

- units that are integral parts of X-ray, high frequency, or electrostatic-coating equipment;
- those used with Class 2 and Class 3 circuits, sign and outline lighting, electric discharge lighting, and power-limited fire-protective signalling circuits; and
- liquid-filled or dry-types used in research, development or testing if other effective protection is provided.

Vaults are required for dry-type, high-fire-point-liquid-insulated and askarel-insulated transformers installed indoors and rated over 35kV; or those transformers that are a fire hazard to employees and are oil-insulated. These vaults are to be constructed so as to contain fires and combustible liquids and prevent unauthorized access. Vault doors must have locks and latches installed so that these doors can be opened from the inside. No foreign duct or pipe may enter or pass through a vault, nor can any unrelated material be stored inside a transformer vault.

The operating voltage of exposed parts on all transformers must be marked by warning signs or other visible markings on the equipment or structure. Where an oil-insulated transformer is attached or adjacent to a building or combustible material, that building and material, along with fire escapes and door and window openings, must be protected from fire initiated at the transformer.

Capacitors must have an automatic means to drain the stored charge when disconnected from their power supply unless they are surge capacitors or those used as components in other equipment. Capacitors rated over 600V must also have disconnects or isolating switches (with no interrupting rating) interlocked to a load-interrupting device, or have highly visible caution signs, to protect against switching the load current.

For those series capacitors rated over 600V and installed after April 16, 1981, proper switching must be made by mechanically sequenced isolating and bypass switches; interlocks; or the proper switching procedures must be prominently displayed at the switching location.

Storage battery installations must have provisions for ventilation to prevent the buildup of explosive gases.

Storage batteries provide backup and communications system power. As the cells generate electricity they release small quantities of explosive gases (depending on how they are constructed). That is why OSHA requires good ventilation where these batteries are kept. What is adequate ventilation is up to the employer to determine using sound environmental engineering, as long as the intent of the rule is met.

This section describes the required safety-related items to be included for the protection of the end-user of specific purpose equipment. These items include a proper disconnecting means, lockable access doors and covers, and warning signs in addition to specific working clearance requirements. As this specific purpose equipment was not adequately covered by the general requirements in OSHA Subpart S, these safety-related items are now included here.

Sec. 1910.306(a)—Electric signs and outline lighting

For electric signs and outline lighting operated by electronic or electromechanical controllers located physically outside of the sign itself, a disconnecting means capable of being locked in the open position must be provided either inside or within sight of the controller. All other signs and outline lighting (except portable signs) must have a disconnect located within sight of the sign or outline lighting. The disconnecting means must be capable of opening all ungrounded circuits simultaneously.

Access doors or covers for access to uninsulated parts of signs or outline lighting with voltages exceeding 600V must have a safety interlock switch which disconnects the primary feeder circuit to the equipment when the door or cover is opened. An approved alternative to the safety interlock switch would be an access door or panel requiring other-than-ordinary tools for opening.

Sec. 1910.306(b)—Cranes and hoists

A readily-accessible disconnecting means must be provided between the runway contact conductors and their power supply. In addition, a second disconnecting means, capable of being locked in the open position, must be provided in the leads from the power pickup on the crane or hoist that will deenergize all power on the unit.

If this additional disconnecting means is not accessible to the operator in the crane or monorail hoist operating station, some other means must be available to allow the operator to remove power to all motors of the crane or

Persons servicing cranes must have a disconnect means accessible to them so that they can remove all power to the equipment.

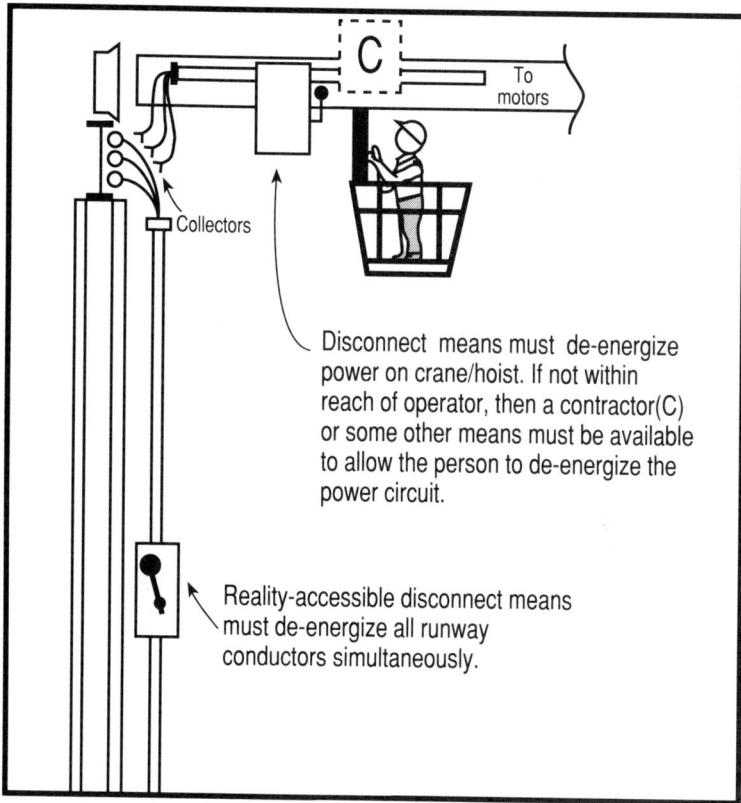

Fig. 6.1. Disconnect requirements for a manned crane or hoist.

hoist from the operating station (**Fig. 6.1**).

The additional disconnecting means is not required if the hoist unit is:

- operated from floor level;
- is within sight of the primary disconnecting means;
- and is not equipped with a fixed work platform for servicing the unit.

A limit switch or other device must be provided to

prevent the hoisting mechanism from traveling above its upper safety limit.

For those live parts of hoists and cranes that may require servicing or inspection while energized, a 2ft-6in. working clearance is required in the direction of access. If these parts are enclosed, the cabinet doors must be capable of being opened 90° or be removable.

Sec. 1910.306(c)—Elevators, dumbwaiters, escalators, and moving walks

A single main disconnecting means is required for each elevator, dumbwaiter, escalator, or moving walk and must be capable of disconnecting all ungrounded power supply conductors simultaneously.

In those multicar installations where interconnections between control panels are necessary and where there is the possibility of energizing from another power source, a warning sign must be mounted on or near the disconnecting means stating : "WARNING — PARTS OF THE CONTROL PANEL ARE NOT DE-ENERGIZED BY THIS SWITCH." This provision applies for those units installed after April 16, 1981.

Where a control panel is not in the same area as the drive machinery, it must be installed in a cabinet with lockable doors.

Sec. 1910.306(d)—Electric welders

For motor-generator, AC-transformer, and DC-rectifier arc welders, a disconnecting means must be provided if the welding equipment is not equipped with an integral

Welding machines are widely used in construction and maintenanace operations and thus are covered by OSHA rules requiring that they be provided with a disconnect means.

disconnecting means.

For resistance welders, a disconnecting means must be provided to isolate the welder or its controller from its power supply. The sizing of this disconnect is determined from the ampacity rating of its supply conductor.

Sec. 1910.306(e)—Data-processing systems

In data processing or computer rooms, separate disconnecting means capable of being controlled from the main exits must be provided to disconnect power to all electronics and computer air conditioning equipment located within the room. This provision is somewhat similar to that of NE Code Sec. 645-10.

Sec. 1910.306(f)—X-ray equipment

For X-ray equipment used in other than medical or dental applications, such as a unit used in an industrial facility to check the quality of parts, a disconnecting means located so as to be readily accessable from the X-ray controls must be provided to isolate the equipment from its power source. A properly rated grounding-type attachment plug and receptacle may be used as this disconnecting means when that equipment is connected to a 30A or less 120V branch circuit.

In those instances where multiple units of equipment are operated from the same high-voltage power circuit, each unit must have its own high-voltage or equivalent disconnecting means constructed so that operators and other personnel cannot come into contact with energized parts.

Radiographic and fluoroscopic equipment must be constructed so that energized current-carrying parts are enclosed to protect operating personnel. Enclosures with safety interlocks that automatically de-energize the equipment when opened satisfy this requirement.

Diffraction- and irradiation-type equipment must have a visual indicating means showing when the equipment is energized, or must be enclosed to prevent accidental contact with live parts. Again, an enclosure with safety interlocks that automatically de-energize the equipment when opened also satisfy this rule.

Sec. 1910.306(g)—Induction and dielectric heating equipment

Induction and dielectric heating equipment and accessories used for other than medical, dental, or appliance applications, must have its converting equipment, including the DC line and high-frequency circuits (but not including output and remote-control circuits) housed in an enclosure made of noncombustible material.

Enclosures housing equipment with voltages of 500 to 1000V AC or DC, access doors must have locks, or have safety interlocks that automatically de-energize the equipment when the doors are opened.

For enclosures housing equipment with voltages over 1000V AC or DC, a mechanical lockout with the disconnecting means, preventing access into the enclosure until the housed equipment is de-energized, must be provided. In lieu of this, safety interlocks and door locks on both doors may be provided to satisfy this requirement.

For enclosures housing equipment with voltages from

250V to 500V AC or DC, "DANGER" labels must be affixed to the enclosure and must be visible even when the enclosure doors are open. Adequate shielding must be provided for the equipment and interlocks must be provided at any access door or panel (unless operating at 150VAC or less, or at DC ground potential) to remove power.

Each unit of heating equipment must be provided with its own disconnecting means capable of isolating the unit from its power source. For units with remote control capability, a selector switch, interlocked so as to prevent powering from multiple points at the same time, must be provided. Foot-activated control switches must have a shield over the switch to prevent accidental operation.

Sec. 1910.306(h)—Electrolytic cells

This section applies to electrolytic cells, electrolytic cell lines (assemblies of electrically interconnected electrolytic cells supplied by a DC power source), and process power supplies used in the production of aluminum, cadmium, chlorine, copper, fluorine, hydrogen peroxide, magnesium, sodium, sodium chlorate, and zinc.

Electrolytic cells are receptacles or vessels in which electromechanical reactions are caused by applying energy for the purpose of refining or producing usable materials. A cell line working zone is the 3-dimensional space in which operation or maintenance activities are normally performed in the vicinity of exposed energized surfaces of cell lines or their attachments.

Exceptions: The overcurrent protection of electrolytic cell DC process power circuits is not required to comply with Sec. 1910.304(e), nor is the grounding of equipment located or used within the cell line working zone or associated with the cell line DC power circuits required to comply with Sec. 1910.304(f). Electolytic cells, cell line conductors, cell line attachments, and the wiring of auxiliary equipment and devices within the cell-line working zone are not required to comply with Sec. 1910.303 (General Requirements). In addition, this equipment and wiring is not required to comply with the requirements listed in Sec. 1910.304(b) "Branch Circuits" and (c) "Outside conductors, 600V, nominal, or less".

Where a cell line has more than one power source, each DC process power supply on the cell line must have a disconnecting means on its load side to isolate it from the cell line circuit. Removable links or removable conductors may be used in lieu of the disconnecting means.

Within the cell line working zone, portable electrical equipment frames and enclosures must not be grounded if the cell line circuit voltage exceeds 200V DC, or if the frames are unguarded. This equipment, however, must be distinctively marked and not be able to be interchangeable with grounded portable electrical equipment.

For those circuits supplying power to ungrounded receptacles for hand-held, cord- and plug-connected equipment, the power to these circuits must be supplied from an isolating transformer. In addition, these circuits must be ungrounded and must be electrically isolated from any distribution system that is supplying areas other than the cell line working zone.

Receptacles and matching plugs used for powering ungrounded equipment must not contain provisions for terminating a grounding conductor. In addition, their configuration should prevent their use on grounded equipment. For those receptacles on circuits supplied by an isolating transformer with an ungrounded secondary, a distinctive configuration along with distinctive markings must be utilized to prevent their use in areas other than the cell line working zone.

Within the cell-line working zone, exposed conductive surfaces of fixed and portable electrical equipment, such as housings, cabinets, boxes, motors, and raceways, along with the AC systems supplying this equipment need not be grounded.

Motors, transducers, sensors, control devices, alarms, and other auxiliary electrical devices mounted on an electrolytic cell or other energized surface can be connected by:

- multiconductor hard-usage or extra-hard-usage flexible cord;
- wire or cable in suitable raceways;
- or exposed metal conduit, cable tray, or armored cable installed with insulating breaks.

Fixed electrical equipment mounted directly on an energized conductive surface of a cell line must be bonded to that surface. This equipment must also be bonded to its attachments and auxiliaries.

Air and water hoses and other nonelectric connections to an electrolytic cell, its attachments, or auxiliary equipment must not be constructed with any continuous conductive reinforcing material such as wire, armor, or braid.

Those cranes and hoists entering a cell-line working zone may have ungrounded conductive surfaces. In addition, those portions of overhead cranes and hoists that come into contact with energized electrolytic cells or energized attachments must be insulated from ground.

Remote controls for cranes and hoists entering a cell line working zone must utilize one or more of the following:

- an insulated and ungrounded control circuit;
- a nonconductive rope operator;
- a pendant pushbutton with a nonconductive supporting means and a nonconductive or ungrounded exposed surface;
- or radio control.

Sec. 1910.306(i)—Electrically driven or controlled irrigation machines

Utilization equipment installed after April 16, 1981 must comply with the following rules.

For electrically driven or controlled irrigation machines with stationary points, a ground rod must be driven at, and connected to, the stationary point for lightning protection.

On center pivot irrigation machines, a readily accessible and lockable main disconnecting means must be provided at the electrical power connection point to the machine. In addition, each motor and controller shall also have a disconnecting means.

Sec. 1910.306(j)—Swimming pools, fountains, and similar installations

The scope of this section applies to electrical wiring for permanently installed or storable swimming, wading, therapeutic (not including health care), and decorative pools and fountains. Also covered is wiring to any metallic auxiliary equipment, such as pumps and filters, and any other equipment in or adjacent to these pools or fountains.

Fig. 6.2 shows the dimensional clearance of areas where specific types of receptacles may be installed at permanently installed pools. The dimensions shown are taken from the inside walls of the pool and represent the shortest path the supply cord of an appliance connected to the receptacle would follow without piercing a floor, wall, or ceiling of a building or other effective permanent barrier.

Fig. 6.3 shows the dimensions required for areas where lighting fixtures and lighting outlets may be installed above and around a permanently installed pool. Note

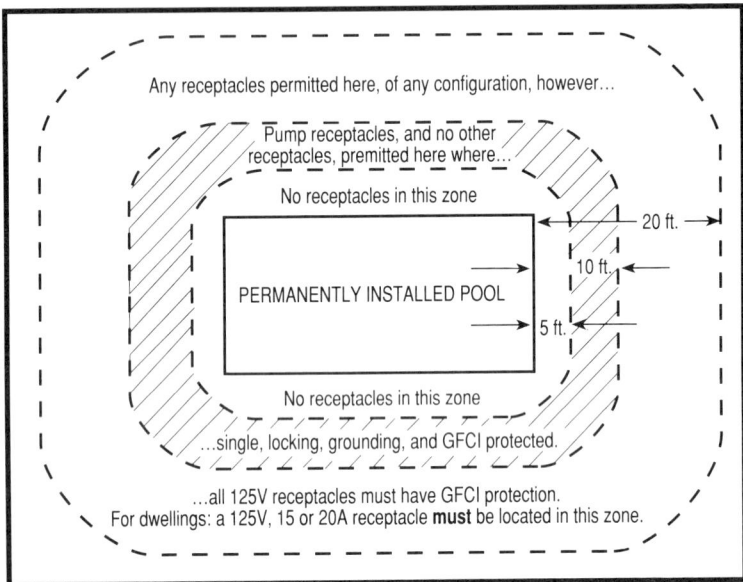

Fig. 6.2. Rules on the installation of receptacles around permanently installed pools.

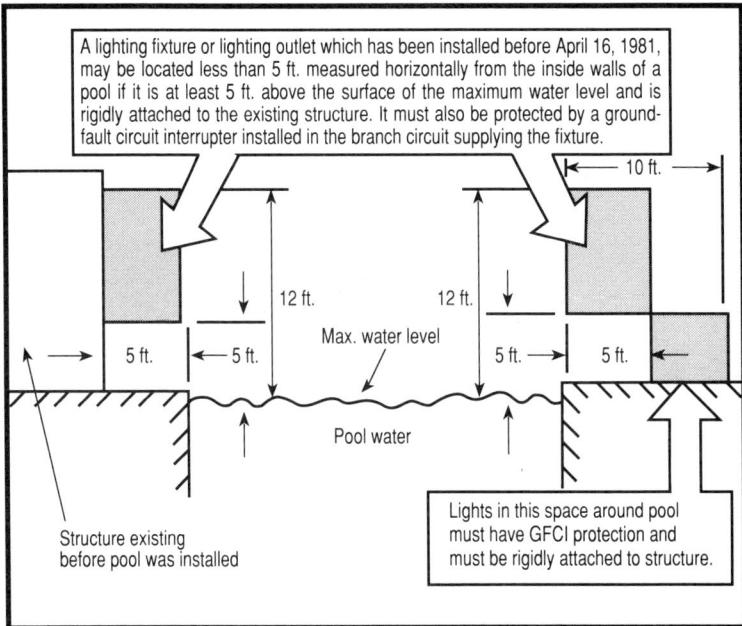

A lighting fixture or lighting outlet which has been installed before April 16, 1981, may be located less than 5 ft. measured horizontally from the inside walls of a pool if it is at least 5 ft. above the surface of the maximum water level and is rigidly attached to the existing structure. It must also be protected by a ground-fault circuit interrupter installed in the branch circuit supplying the fixture.

10 ft.

12 ft.

12 ft.

Max. water level

5 ft.

5 ft.

5 ft.

5 ft.

Pool water

Lights in this space around pool must have GFCI protection and must be rigidly attached to structure.

Structure existing before pool was installed

Fig. 6.3. Rules on the installation of lighting fixtures and outlets around permanently installed pools.

that GFCI protection is required on all circuits feeding these fixtures and outlets.

Flexible cords feeding cord- and plug-connected lighting fixtures installed within 16 ft of the water surface, and fixed and stationary equipment used with permanently installed pools, cannot exceed 3 ft in length and must have a copper equipment grounding conductor with a grounding-type attachment plug.

For all underwater fixtures operating at more than 15V, a ground-fault circuit interrupter must be installed on the branch circuit feeding these fixtures. In addition, no underwater lighting fixtures operating at voltages over 150V between conductors may be installed.

Electric equipment operating at more than 15V to ground, including power supply cords, must be protected by a ground-fault circuit interrupter. This rule applies to new as well as existing equipment installed after April 16, 1981.

Swimming pools are also workplaces if you or your client hires employees to maintain and supervise the facilities.

Nowhere is the need for safety procedures greater than in areas that potentially can contain a flammable atmosphere. The NE Code sets down rules in Article 500 that must be followed in such locations. Similarly, OSHA has laid down a set of regulations that specifically apply to hazardous locations.

Sec. 1910.307(a) — Scope

The regulations cover electrical installations located where there is a chance of fire or explosion because of flammable liquids, vapors, gases, dusts, and other material that could be present, and a likelihood that a flammable or combustable concentation could be present. Each section, area or room must be considered individually when determining its classification.

OSHA follows the NE Code in dividing hazardous locations into six designations. The definition of the hazards involved in each are given in Sec. 1910.399.

Class I, Division 1 locations are those in which:

• hazardous concentrations of flammble gases or vapors may exist under normal operating conditions;

• or hazardous concentrations of the gases or vapors may exist frequently because of repair, maintenance operations, or leakage;

• or where breakdown or faulty operation of equipment or processes might release hazardous concentrations of flammable gases or vapors, and simultaneously may cause failure or electrical equipment.

Hazardous (classified) locations are found in many industries in which flammable vapors and gases are processed. OSHA regulations require the use of equipment that has been found suitable by an NRTL for the atmosphere involved.

Locations that meet this classification can usually be found where volatile flammable liquids or liquified flammable gases are transferred from one container to another. Other examples include:

• interiors of spray booths and areas near spraying and painting operations where volatile flammable solvents are employed;

• locations containing open tanks or vats of volatile flammable liquids;

• drying rooms or compartments for the evaporation of

flammable solvents;

• locations containing fat and oil extraction equipment using volatile flammable solvents;

• portions of cleaning and dyeing plants where flammable liquids are used;

• gas generator rooms and other portions of gas manufacturing plants where flammable gas may escape;

• inadequately ventilated pump rooms for flammable gas or liquids;

• the interiors of refrigerators and freezers in which volatile flammable materials are stored in open, lightly stoppered, or easily ruptured containers;

• and all other locations where ignitible concentrations of flammable vapors or gases are likely to occur in the course or normal operations.

Class I, Division 2 locations are those:

• in which volatile flammable liquids or gases are handled, processed, or used but will normally be confined within closed containers or closed systems from which they can escape only in case of accidental rupture or breakdown of the containers or systems, or due to abnormal operation of equipment;

• or in which hazardous concentrations of these gases or vapors are normally prevented by positive mechanical ventilation, and which might become hazardous through failure or abnormal operations of the ventilating equipment;

• or areas adjacent to a Class I, Div. 1 location to which hazardous concentrations of gases or vapors might occasionally be communicated (unless such communication is prevented by adequate positive-pressure ventilation from a source of clean air, and effective safeguards against ventilation failure are provided).

Locations meeting these criteria includes those where volatile flammable liquids, gases, or vapors are used, but which would become hazardous only in case of an accident or some unusual operating condition. The quantity of flammable material that might escape in case of accident, the adequacy of ventilating equipment, the total area involved, and the record of the industry with respect to explosions or fires are all factors that merit consideration in determining the classification in such cases. Piping without valves, checks, meters, and similar devices do not ordinarily introduce a hazardous condition even though

Specific information is required to be displayed on explosionproof equipment so that persons installing a piece of electrical equipment can readily identify whether it is suitable for the location.

used for flammable liquids or gases.

Locations used for storage of flammable liquids, or liquified or compressed gases in sealed containers are not normally considered hazardous unless also subject to other hazardous conditions. On the other hand, electrical conduits and their associated enclosures separated from process fluids by a single seal or barrier are classified as a Div. 2 location if the outside of the conduit and enclosures is a nonhazardous location.

Class II, Division 1 locations are those in which:

• combustible dust is or may be in suspension in the air under normal operation conditions in quantities sufficient to produce explosive or ignitible mixtures;

• or mechanical failure or abnormal operation of machinery or equipment might cause such explosive or ignitible mixtures to be produced, and might provide a source of ignition through simultaneous failure of electrical equipment, operation of protective devices, or from other causes;

• or where combustible dusts of an electrically conductive nature may be present.

This classification may include areas of grain handling and processing plants, starch plants, sugar-pulverizing plants, malting plants, hay-grinding plants, coal pulverizing plants, areas where metal dusts and powders are produced or processed, and other similar locations that contain dust-producing machinery and equipment (except where the equipment is dusttight or vented to the outside). Combustible dusts that are electrically nonconductive include those produced in the handling and processing of grain and grain products, pulverized sugar and cocoa, dried egg and milk powders, pulverized spices, starch and pastes, potato and wood flower, oil meal from beans and seed, dried hay, and other organic materials. Electrically conductive dusts such as those containing magnesium or aluminum are particularly hazardous and the use of extreme caution is necessary to avoid ignition and explosion.

Class II, Division 2 is a location in which:

• combustible dust will not normally be in suspension in the air in quantities sufficient to produce explosive or ignitable mixtures, and dust accumulations are normally insufficient to interfere with the normal operation or electrical equipment or other apparatus;

• or where dust may be in suspension in the air as a result of infrequent malfunctions of handling or processing equipment, and dust accumulations resulting therefrom may be ignited by abnormal operation or failure of electrical equipment or other apparatus.

This classification includes locations where dangerous concentrations of suspended dust are not likely to occur but where dust accumulations might form on or in the vicinity of electrical equipment. These areas may contain equipment from which appreciable quantities of dust could escape under abnormal operating conditions, or be adjacent to a Class II, Div. 1 location from which dust may be communicated and put into suspension under abnormal operating conditions.

Class III, Division 1 locations are those in which easily ignitible fibers or materials producing combustible flyings are handled, manufactured, or used.

Such locations usually include some parts of textile mills, combustible fiber manufacturing and processing plants, cotton gins and cotton-seed mills, flax-processing plants, clothing manufacturing plants, woodworking plants, and industries involving similar hazardous processes or conditions. Easily ignitible fibers and flyings include rayon, cotton (including cotton linters and cotton waste), sisal or henaquen, istle, jute, hemp, tow, cocoa fiber, oakum, baled waste kapok, Spanish moss, excelsior, and other materials of similar nature.

Class III, Division 2 locations are those in which easily ignitible fibers are stored or handled (except in process of manufacture).

Sec. 1910.307(b) — Electrical installations

Electrical equipment and wiring methods used in hazardous locations must be of a type that is assures the safety of those working in the area. Several general categories of equipment are listed by the OSHA regulations as being acceptable.

• Equipment that has been approved for use in the specific location, including having a maximum surface temperature that is lower than the ignition temperature of the hazardous material or vapors within the area. Suitability of equipment or materials for a specific purpose, environment, or application may be determined by a nationally recognized testing laboratory (NRTL), inspection agency or other organization concerned with product evaluation as part of its "listing" and "labeling" program. After an NRTL tests a piece of equipment for safety within a certain environment, a label is attached by the manufacturer to the item. It includes the identification of the NRTL and other pertinent data.

Another way of gauging acceptability for use in a particular hazardous location is by the equipment's listing in an NRTL document that states that such equipment meets nationally recognized standards or has been tested and found safe for use in a specified manner.

• Equipment and associated wiring that has been approved as "intrinsically safe" is permitted within any hazardous location for which it is approved. This type of equipment has been tested by an NRTL to standards that determine that the components and/or system is incapable or releasing sufficient energy to cause ignition of a hazardous atmosphere.

• Equipment that is considered to be safe for the location is allowed. This provision permits employers to use equipment and techniques that they can demonstrate

Factory-sealed devices have adequate internal provisions to prevent an internal explosion from being communicated to the surrounding atmosphere without the necessity of installing an external seal fitting in the raceway to the unit.

will offer protection from the hazards arising from the materials used on the site. OSHA references the provisions of Chapter 5 of the NE Code for guidelines on safe practices in hazardous (classified) locations.

Regulations of this section lay down specific requirement for the marking of equipment that is used in a hazardous location. The class and operating temperature or range (based on a 40°C ambient) for which it is approved must be shown. The "group" for which it is suitable must also be shown.

For the purpose of testing, approval, and area classification, NE Code Sec. 500-3 FPNs assembles various materials that can be considered as hazardous within the intent of the article and assembles them into atmospheric groups A through G posing basically the same degree of hazard. Some examples of the types of hazardous materials in each group is included, but FPN No. 12 refers the reader to NFPA documents 325M and 497M for more specific details. It is then the responsibility of the persons designing the facility to make sure that the temperature

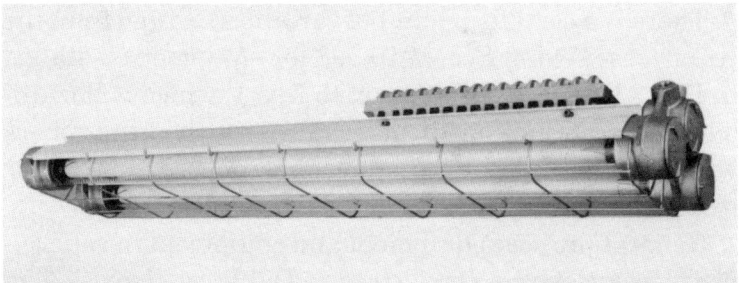

Explosionproof lighting fixtures such as this fluorescent one, are required to have a surface temperature limit below that of the ignition temperature of the hazardous vapor or gas that potentially can be encountered in the area.

markings do not exceed the ignition temperature of the specific hazardous material that will be encountered.

Non heat-producing equipment such as conduit, fittings, and junction boxes, and heat-producing equipment whose maximum temperature is no more than 100°C (212°F) are exempt from this temperature-marking rule. Several other types of equipment have some degree of exemptions from the marking rule.

Sec. 1910.307(c) — Conduits

In a hazardous location, conduits must be threaded and and made up wrench-tight. Where it is not practical to make up a wrench-tight connection, a bonding jumper must be used. The purpose here is to ensure electrical continuity. Arcing across loose joints could occur if a phase-to-ground fault exists, thus triggering an explosion within the area.

Sec. 1910.307(d) — Equipment in Div. 2 locations

There is a certain amount of variation allowed from the rules presented in Sec. 1910.308 for equipment installed in Div. 2 locations. In contrast to Div. 1, which is hazardous under normal operating conditions, the atmosphere within a Div. 2 location is not hazardous except under unusual circumstances.

General-purpose equipment and equipment in general-purpose enclosures are allowed in Div. 2 locations as long as the equipment does not constitute a source of ignition under normal operating conditions. One of the most often cited examples of this is a general-purpose junction box containing a terminal block. The wires terminating on the block rarely pose an arcing threat, and the atmosphere within the enclosure rarely would be in the explosive range. It would take a most unusual occurrence for the installation to pose a threat.

This regulation, however, also confirms that an equipment upgrade is permitted in a Div. 2 location. For instance, Div. 1 equipment can be used in Div. 2 locations as long as it meets the class and group requirements of the area in which it is installed.

Systems over 600V have to meet a specific set of additional OSHA rules.

The rules in this chapter apply to systems that are not completely addressed or not covered at all by the rules set forth in Secs. 1910.301 through 1910.307. For instance, most of the previous regulations apply to systems operating at voltages of 600V or lower. References are made in several places to how these apply to systems of over 600V, and give some additional rules for these systems. In this section, some specific rules that apply only to these higher-voltage systems are listed. Likewise, rules that apply to emergency-power systems, communications systems, and others are addressed.

Sec. 1910.308(a) — Systems over 600V nominal

Above-ground conductors of systems operating at over 600V can be run in rigid metal conduit, IMC, cable trays, in cable bus, as MC cable listed for the purpose, or other suitable raceways. However, nonmetallic cable runs, bare conductors, or busway must be installed where accessible only to qualified persons. Again, the definition in Sec. 1910.399 of a qualified person is, "one familiar with the construction and operation of the equipment and the hazards involved. Open runs of insulated wires or cables must be supported so as to prevent physical damage to the sheath or covering.

Conductors emerging from the ground must be enclosed in approved raceways. This rule applies to runs installed after April 16, 1981.

A means must be provided to completely electrically

Circuit protective devices utilized in this system operating at over 600V are enclosed in metalclad gear to comply with the OSHA regulations.

isolate equipment so that it can be inspected and repaired. If the isolation means does not have sufficient interrupting capacity for the load current, it must be interlocked with an approved circuit interrupter or provided with a sign warning against opening under load. Fused cutouts within a building or vault must be approved for the purpose and readily accessible for fuse replacement.

All energized switching and control devices must be enclosed in effectively-grounded metal cabinets or enclosures. The cabinets or enclosures must be locked so that only authorized qualified persons have access, and they must be marked with a sign warning of the presence of energized parts.

Circuit breakers of systems over 600V must be metal enclosed or in fire-resistant switchgear, and have a method to clearly indicate the open and closed positions. When mounted in a room accessible only to qualified personnel, open mounting of the circuit breakers is permitted. Circuit breakers and other protective equipment must be operable without having to open the locked doors of their enclosure.

Mobile and portable equipment such as substations, shovels, draglines, hoists, dredges, conveyors, etc. operating at over 600V must meet specific requirements. Multiconductor portable cable may supply mobile equipment. An equipment grounding conductor, either insulated or bare, must be run with circuit conductors inside the multiconductor cable jacket. A metallic enclosure must be provided on the mobile machine for enclosing the terminals of the power-supply cable. It must contain provisions for the solid connection of the ground wire for effectively grounding the machine frame. Provisions must be made to prevent any strain or pull on the cable from stressing the cable connections. The enclosure must be locked so that access is possible only for qualified persons and must have a sign warning of the presence of energized parts. The collector rings of revolving type shovels and similar machines must be guarded.

In tunnels, the higher-voltage conductors must be installed in either: metal conduit or other metal raceway; as MC cable; or in other approved multiconductor cable. An equipment grounding conductor, either bare or insulated, must be run with the circuit conductors in the cable or raceway. Bare terminals of transformers, switches, motor controllers, and other electrical equipment must be enclosed to prevent accidental contact with live parts. The enclosures must be dripproof, weatherproof, or submersible, as required by the conditions of application. A disconnect switch is required at transformers and motors to open all ungrounded conductors to the equipment. All noncurrent-carrying metal parts of electrical equipment, metal raceways and cable sheaths must be effectively grounded and bonded to all metal pipes and rails at the portal and at intervals not exceeding 1000 ft throughout the tunnel.

Sec. 1910.308(b) — Emergency power systems

The provisions of this subsection apply to circuits, systems, and equipment intended to supply power for illumination and special loads in the event of failure of the normal supply.

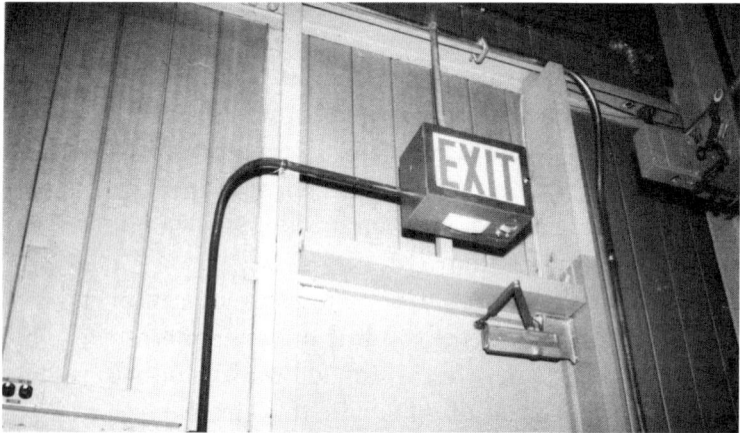

Exit sign and other emergency circuits must be run in separate raceways from power and other circuits in order to safeguard them as much as possible.

Emergency circuit wiring must be kept entirely independent of all other wiring and equipment and must not enter the same raceway, cable, box, or cabinet except:

- where common circuit elements suitable for the purpose are required;
- or for transferring power from the normal to the emergency source.

Emergency lighting systems must be so arranged that the failure of any individual lighting element, or burning out of a lamp, will not leave any space in total darkness.

Sec. 1910.308(c) — Class 1, 2, and 3 circuits

Definitions are given of circuits characterized by their usage and electrical power limitation that differentiates them from light and power circuits. They are classified in accordance with their respective voltage and power limitations.

Class 1 power-limited circuit is supplied from a source having a rated output of not more than 30V and 1000VA.

Class 1 remote control or signaling circuit has a voltage that does not exceed 600V, but the power output of the source is not limited.

Class 2 circuit is limited either inherently (not due to overcurrent protection) or by a combination of a power source and overcurrent protection. The maximum circuit voltage is 150V AC or DC for an inherently-limited power source, and 30V AC or DC for a source limited by overcurrent protection.

Class 3 circuit is also limited either inherently or by a combination of a power source and overcurrent protection.

Power-limited wiring is defined in the regulations as either Class 1, Class 2, or Class 3 circuits depending upon factors as their energy levels.

The maximum circuit voltage is 100V AC or DC for an inherently-limited power source, and 150V AC or DC for a power source limited by overcurrent protection.

The voltage limitations for all of the circuits is based upon a sinusoidal AC or continuous DC source being applied.

A Class 2 or Class 3 power supply unit must be durably marked where plainly visible with the class of supply and its electrical rating. This provision applies to those units installed after April 16, 1981.

Sec. 1910.308(d) — Fire protective signaling systems

The provisions of this subsection applies to systems installed after April 16, 1981. Fire protective signaling systems must be classified as being either power limited or non-power limited. Depending upon the classification, a different set of rules apply.

Power limited circuit power sources can be either inherently limited (not due to overcurrent device) or limited by a combination of a power source and overcurrent protection. Where open conductors are installed, the circuits must be separated by at least 2 in. from conductors of light, power, Class 1, and non-power limited fire protective signaling circuits (unless an equally protective method of conductor separation is employed). Conductors of two or more power-limited fire protective signaling circuits or

Fire-protective circuits must be catagorized as being power limited or nonpower limited. Different rules apply, depending upon their category.

Class 3 circuits are permitted in the same cable, enclosure, or raceway. Conductors of one or more Class 2 circuits are permitted within the same cable, enclosure, or raceway with conductors of power-limited fire protective signaling circuits if the insulation of the Class 2 circuit conductors is at least that needed for the power-limited fire protective signaling circuits.

Non-power-limited circuit power supply must have an output voltage not exceeding 600V. These circuits and Class 1 circuits may occupy the same enclosure, cable, or raceway provided all conductors are insulated for the maximum voltage of any of the conductor. Power supply and fire protective signaling circuit conductors are permitted in the same enclosure, cable, or raceway only if connected to the same equipment.

Fire protective signaling circuits must be identified at terminal and junction locations in a manner that will prevent unintentional interference with the signaling circuit during testing and servicing. Power-limited circuits must be durably marked as such where plainly visible at terminations.

Sec. 1910.308(e) — Communications systems

These provisions apply to central-station-connected and non-central-station-connected telephone circuits, radio and TV receiving and transmitting equipment, community antenna TV and radio distribution systems, telegraph, district messenger, outside wiring for fire and burglar alarm, and similar systems. Communication systems do not have to comply with most of the Sec. 1910.303 through 1910.308(d) regulations, except for:

Communications conductors that possibly may come in contact with power circuits over 300V must be provided with a protector approved for the purpose.

- the climbing space requirement of conductors installed outdoors on poles given in Sec. 1910.304(c);
- and the rules governing electrical installations in hazardous (classified) locations given in Sec. 1910.307(b).

Communications conductors that can accidentally come in contact with light or power conductors operating at over 300V must be provided with a protector approved for the purpose. When run on poles, the communication conductors should, where practical, be located below the light and power conductors. They are not allowed to be attached to a crossarm that carries light or power conductors.

A lead-in conductor from an antenna must be provided with an antenna discharge unit or other suitable means to drain static charges. Lead-in or aerial-drop communication cables attached to buildings should be installed so as to avoid the possibility of accidental contact with electric light and power conductors. The clearance between the lead-in conductors and any lightning protection conductors must be at least 6 ft. If exposed to contact with electric light and power conductors, the metal sheath of aerial communications cables entering buildings must be grounded or interrupted close to the entrance to the building by an insulating joint or equivalent device. Where protective devices are used, they must be grounded in an approved manner.

Outdoor metal structures supporting antennas and self-supporting antennas must be located as far away from overhead conductors of electric light and power circuits over 150V to ground as necessary to avoid the possibility of the antenna or structure falling into or making accidental contact with such circuits. Masts and metal structures supporting antennas must be permanently and effectively grounded, and the grounding conductor run must be continuous without a splice.

Transmitters must be enclosed in a metal frame or grill, or separated from the operating space by a barrier. All of the metallic parts must be effectively connected to ground. All external metal handles and controls accessible to the operating personnel must be effectively grounded. Unpowered equipment and enclosures are considered grounded where connected to an attached coaxial cable with an effectively grounded metallic shield.

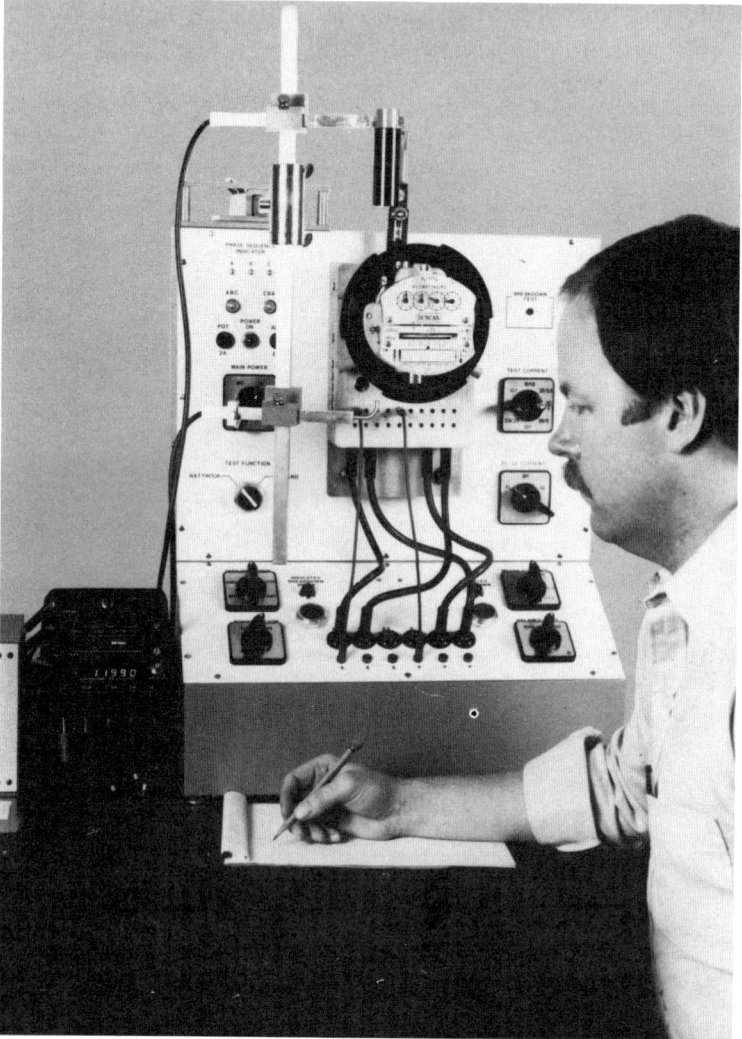

Testing of electrical equipment is required in order to assure its safety when placed in operation. NRTLs perform this important function.

SEC. 1910.7 – NATIONALLY-RECOGNIZED TESTING LABORATORIES

A nationally-recognized testing laboratory (NRTL) is an OSHA-accredited organization that safety-tests, accepts, lists, or labels products to be used in the workplace.

The NRTL definition has undergone the single most revolutionary change since the Part I regulation was published in 1981. The accreditation reference is totally new, and so is the absence in Sec. 1910.399 (Definitions) of the phrase, "such as, but not limited to Underwriters Laboratories, Inc. and Factory Mutual Engineering Corp."

OSHA scrapped all references to UL and FM and implemented in 1988 an accreditation program for labs who want NRTL recognition. The agency was forced to make these changes by a lawsuit brought about MET Electrical Testing, Baltimore, Md. MET became the first lab to earn OSHA approval in 1989. While OSHA already had an NRTL accreditation program on its books (29 CFR Part 1907), it was never put in place since it became official in 1973. The program has now been incorporated in 29 CFR Part 1910, which is OSHA's general industry safety standards, including electrical safety rules Parts I-IV.

Part 1907 is now Sec. 1910.7, and Sec. 1910.399(a) now asks reader to refer to Sec. 1910.7 for the NRTL definition. This wording applies to all OSHA electrical rules, including Subpart K that became law in 1986.

All equipment and installations in the workplace must be, according to Sec. 1910.399(a), accepted, or certified, or listed, or labeled, or otherwise determined to be safe by a nationally-recognized testing laboratory.

There are exceptions given to this rule:

• Equipment or installations that are not NRTL-approved but which are inspected or tested by another federal agency, or by state, or local authorities in accordance with the National Electrical Code.

• Custom-installed equipment whose manufacturer-supplied test data must be kept on hand by the employer for OSHA inspection.

• If the equipment or installation is acceptable to the Assistant Secretary of Labor and approved in accordance with the regulation.

OSHA defines *accepted* as an installation that is inspected and found by an NRTL to conform to specified plans or to procedures of applicable codes.

Certified applies to a piece of equipment if it:

• has been tested by an NRTL to meet nationally-recognized standards or to be safe to use in a specific manner;

• or is or a kind whose production is periodically inspected by an NRTL;

• and it bears a label, tag, or other record of certification.

Equipment is *labeled* if there is attached to it a label, symbol, or other identifying mark of an NRTL which:

• makes periodic inspections of the production of such equipment;

• and whose labeling indicates compliance with nationally recognized standards or tests to determine safe use in a specified manner.

Equipment is *listed* if it is of a kind mentioned in a list which:

• is published by a nationally recognized laboratory

Nationally Recognized Testing Laboratory	LISTING NO.
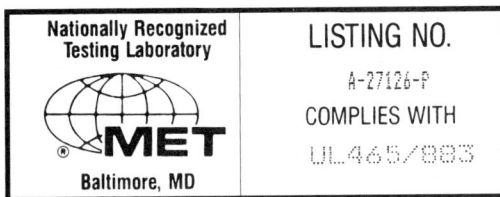 **MET** Baltimore, MD	A-27126-P COMPLIES WITH UL 465/883

NRTLs must have the proper testing equipment, calibration, facilities, trained staff, procedures, and quality control needed to properly examine and test equipment and materials.

---

that makes periodic inspection of the production of such equipment;

• and states such equipment meets nationally recognized standards or has been tested and found safe for use in a specific manner.

NRTLs are approved to accept, label, list, or test for each specified piece of equipment. They test for a standard that is "recognized in the United States as a safety standard providing an adequate level of safety" as long as it is compatible and current with national codes (like the NEC) and installation standards, and is developed by a consensus standards organization. The NRTL can also test to a current American National Standards Institute (ANSI) or American Society for Testing and Materials (ASTM) standard. Any other standard the NRTL wishes to test to must be evaluated for safety by the Assistant Secretary of Labor.

There is no "blanket" NRTL recognition for all standards. Each lab must specify the standards it wants to test for in its application. OSHA has, however, issued temporary blanket accreditation to UL and FM that is effective to July, 1993. According to Sec.1910.7, each lab must:

- have the capability, including proper testing equipment, calibration, facilities, trained staff, procedures, and quality control to examine and test equipment and materials;
- use control procedures to identify listed and labeled equipment or materials;
- have the ability to perform and evaluate follow-up inspections on factories where the goods are made;
- be able to conduct field inspections to check on proper use of its mark on products;
- be independent of employers buying the tested product and of the manufacturers and distributors who supply them or their equipment or materials;
- maintain accurate, effective, and unbiased findings and reports;
- and handle and respond to complaints and disputes fairly and reasonably.

OSHA requires labs who want to become NRTLs to show that they are competent enough to undertake the tasks. The regulations that outline the recognition process are contained in Appendix A to Sec. 1910.7. Each lab must also specify the testing scope and methods, unless the methods are set out in the test standard; the agency could ask the applying lab to prove the effectiveness of the method.

Foreign labs face an additional criteria; reciprocity. OSHA has to taken into account how and whether the parent country recognizes equipment acceptances, labeling, listings, and test data from U.S. NRTLs, or if the parent country has any such requirements at all.

Once a lab passes the eligibility requirements it is subject to on-site review by qualified technical experts based on "appropriate national consensus standards or international guides with such additions, changes, or deletions as may be necessary..." These experts could also include non-Federal individuals such as OSHA-acceptable consultants or contractors.

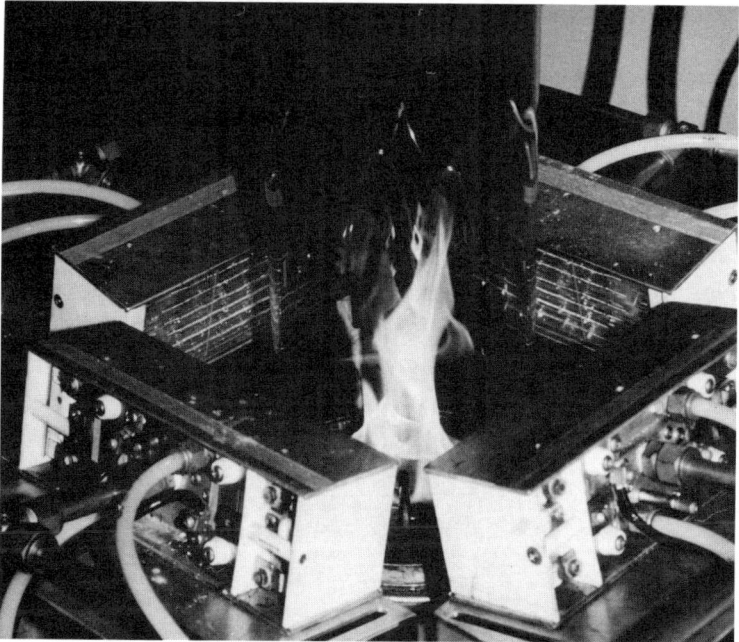

Flame tests are a common procedure carried out at testing laboratories to see if a product is safe for the workplace. As much as possible, NRTLs try to replicate real-life conditions that the product or equipment will be subjected to when in use.

Tray cable must pass rigerous flame tests at an NRTL to assure that when it is applied in a specified manner it meets all requirements for safety.

The application then faces staff review, preliminary findings, public review and comments, and final action. Negative public comments could prompt further a special review of any problems identified in the lab's proposal.

Each NRTL recognition is good for 5 years. The accreditation can be terminated either voluntarily or it can be revoked by the Assistant Secretary of Labor.

This appendix contains the actual OSHA regulations text. These are the words written to be enforced by OSHA inspectors and backed by the courts. This is why the language contained in here is different than that in the NFPA 70E Part I standard from which these regulations are derived. The only difference between what is shown here and in the rules as they appear in the Federal Register is that the Section and Subsection titles have been made bold to permit easy correlation with the explanation of the rules in the front of this book.

Subpart S — Electrical

Authority: Secs. 6, 8, Occupational Safety and Health Act of 1970 (29 U.S.C. 655, 657); Secretary of Labor's Order No. 8 - 76 (41 FR 35736) or 9 - 83 (48 FR 35736), as applicable; 29 CFR Part 1911.

Source: 46 FR 4056, Jan. 16, 1981, unless otherwise noted.

General

Sec. 1910.301 Introduction.

This subpart addresses electrical safety requirements that are necessary for the practical safeguarding of employees in their workplaces and is divided into four major divisions as follows:

(a) *Design safety standards for electrical systems.* These regulations are contained in 1910.302 through 1910.330. Sections 1910.302 through 1910.308 contain design safety standards for electric utilization systems. Included in this category are all electric equipment and installations used to provide electric power and light for employee workplaces. Sections 1910.309 through 1910.330 are reserved for possible future design safety standards for other electrical systems.

(b) *Safety-related work practices.* These regulations will be contained in 1910.331 through 1910.360.

(c) *Safety-related maintenance requirements.* These regulations will be contained in 1910.361 through 1910.380.

(d) *Safety requirements for special equipment.* These regulations will be contained in 1910.381 through 1910.398.

(e) *Definitions.* Definitions applicable to each division are contained in 1910.399.

[46 FR 4056, Jan. 16, 1982; 46 FR 40185, Aug. 7, 1981]

Design Safety Standards for Electrical Systems.

Sec. 1910.302 Electric utilization systems.

Sections 1910.302 through 1910.308 contain design safety standards for electric utilization systems.

(a) *Scope* — (1) *Covered.* The provisions of 1910.302 through 1910.308 of this subpart cover electrical installations and utilization equipment installed or used within or on buildings, structures, and other premises including:

(i)   Yards,

(ii)   Carnivals,

(iii) Parking and other lots,

(iv)   Mobile homes,

(v)   Recreational vehicles,

(vi)   Industrial substations,

(vii) Conductors that connect the installations to a supply of electricity, and

(viii) Other outside conductors on the premises.

(2) *Not covered.* The provisions of 1910.302 through 1910.308 of this subpart do not cover:

(i)   Installations in ships, watercraft, railway rolling stock, aircraft, or automotive vehicles other than mobile homes and recreational vehicles.

(ii)   Installations underground in mines.

(iii) Installations of railways for generation, transformation, transmission, or distribution of power used exclusively for operation of rolling stock or installations used exclusively for signaling and communication purposes.

(iv)   Installations of communication equipment under the exclusive control of communication utilities, located outdoors or in building spaces used exclusively for such installations.

(v)   Installations under the exclusive control of electric utilities for the purpose of communication or metering; or for the generation, control, transformation, transmission, and distribution of electric energy located in buildings used exclusively by utilities for such purposes or located outdoors on property owned or leased by the utility or on public highways, streets, roads, etc., or outdoors by established rights on private property.

(b) *Extent of application.* (1) The requirements contained in the sections listed below shall apply to all electrical installations and

106

utilization equipment, regardless of when they were designed or installed.

Sections:

1910.303(b) .... Examination, installation, and use of equipment.

1910.303(c) .... Splices.

1910.303(d) .... Arcing parts.

1910.303(e) .... Marking.

1910.303(f) .... Identification of disconnecting means.

1910.303(g)(2) .... Guarding of live parts.

1910.304(e)(l)(i) .... Protection of conductors and equipment.

1910.304(e)(l)(iv) .... Location in or on premises.

1910.304(e)(l)(v) .... Arcing or suddenly moving parts.

1910.304(f)(l)(ii) .... 2-Wire DC systems to be grounded:

1910.304(f)(l)(iii) and

1910.304(f)(l)(iv) .... AC Systems to be grounded.

1910.304(f)(l)(v) .... AC Systems 50 to 1000 volts not required to be grounded.

1910.304(f)(3) .... Grounding connections.

1910.304(f)(4) .... Grounding path.

1910.304(f)(5)(iv)(a) through

1910.304(f)(5)(iv)(d) ... Fixed equipment required to be grounded.

1910.304(f)(5)(v) .... Grounding of equipment connected by cord and plug.

1910.304(f)(5)(vi) .... Grounding of nonelectrical equipment.

1910.304(f)(6)(i) .... Methods of grounding fixed equipment.

1910.305(g)(l)(i) and

1910.305(g)(1)(ii) .... Flexible cords and cables, uses.

1910.305(g)(l)(iii) .... Flexible cords and cables prohibited.

1910.305(g)(2)(ii) .... Flexible cords and cables, splices.

1910.305(g)(2)(iii) .... Pull at joints and terminals of flexible cords and cables.

1910.307 .... Hazardous (classified) locations.

(2) Every electric utilization system and all utilization equipment installed after March 15, 1972, and every major replacement, modification, repair, or rehabilitation, after March 15, 1972, of any part of any electric utilization system or utilization equipment installed before March 15, 1972, shall comply with the provisions of 1910.302 through 1910.308.

Note: "Major replacements, modifications, repairs, or rehabilitations" include work similar to that involved when a new building or facility is built, a new wing is added, or an entire floor is renovated.

(3) The following provisions apply to electric utilization systems and

utilization equipment installed after April 16, 1981:

1910.303(h)(4) (i)
and (ii) .... Entrance and access to workspace (over 600 volts).
1910.304(e)(1)(vi)(b) .. Circuit breakers operated vertically.
1910.304(e)(1)(vi)(c) .. Circuit breakers used as switches.
1910.304(f)(7)(ii) .... Grounding of systems of 1000 volts or more supplying portable or mobile equipment.
1910.305(j)(6)(ii)(b) .... Switching series capacitors over 600 volts.
1910.306(c)(2) .... Warning signs for elevators and escalators.
1910.306(i) .... Electrically controlled irrigation machines.
1910.306(j)(5) .... Ground-fault circuit interrupters for fountains.
1910.308(a)(1)(ii) .... Physical protection of conductors over 600 volts.
1910.308(c)(2) .... Marking of Class 2 and Class 3 power supplies.
1910.308(d) .... Fire protective signaling circuits.

[46 FR 4056, Jan. 16, 1981; 46 FR 40185, Aug. 7, 1981]

Sec. 1910.303 General requirements.

(a) *Approval.* The conductors and equipment required or permitted by this subpart shall be acceptable only if approved.

(b) *Examination, installation, and use of equipment* — (1) *Examination.* Electrical equipment shall be free from recognized hazards that are likely to cause death or serious physical harm to employees. Safety of equipment shall be determined using the following considerations:

(i) Suitability for installation and use in conformity with the provisions of this subpart. Suitability of equipment for an identified purpose may be evidenced by listing or labeling for that identified purpose.

(ii) Mechanical strength and durability, including, for parts designed to enclose and protect other equipment, the adequacy of the protection thus provided.

(iii) Electrical insulation.

(iv) Heating effects under conditions of use.

(v) Arcing effects.

(vi) Classification by type, size, voltage, current capacity, specific use.

(vii) Other factors which contribute to the practical safeguarding of employees using or likely to come in contact with the equipment.

(2) *Installation and use.* Listed or labeled equipment shall be used or installed in accordance with any instructions included in the listing or labeling.

(c) *Splices.* Conductors shall be spliced or joined with splicing devices suitable for the use or by brazing, welding, or soldering with a fusible metal or alloy. Soldered splices shall first be so spliced or joined as to be mechanically and electrically secure without solder and then soldered.

All splices and joints and the free ends of conductors shall be covered with an insulation equivalent to that of the conductors or with an insulating device suitable for the purpose.

(d) *Arcing parts.* Parts of electric equipment which in ordinary operation produce arcs, sparks, flames, or molten metal shall be enclosed or separated and isolated from all combustible material.

(e) *Marking.* Electrical equipment may not be used unless the manufacturer's name, trademark, or other descriptive marking by which the organization responsible for the product may be identified is placed on the equipment. Other markings shall be provided giving voltage, current, wattage, or other ratings as necessary. The marking shall be of sufficient durability to withstand the environment involved.

(f) *Identification of disconnecting means and circuits.* Each disconnecting means required by this subpart for motors and appliances shall be legibly marked to indicate its purpose, unless located and arranged so the purpose is evident. Each service, feeder, and branch circuit, at its disconnecting means or overcurrent device, shall be legibly marked to indicate its purpose, unless located and arranged so the purpose is evident. These markings shall be of sufficient durability to withstand the environment involved.

(g) *600 Volts, nominal, or less* — (1) *Working space about electric equipment.* Sufficient access and working space shall be provided and maintained about all electric equipment to permit ready and safe operation and maintenance of such equipment.

(i) *Working clearances.* Except as required or permitted elsewhere in this subpart, the dimension of the working space in the direction of access to live parts operating at 600 volts or less and likely to require examination, adjustment, servicing, or maintenance while alive may not be less than indicated in Table S - 1 (on next page). In addition to the dimensions shown in Table S - 1, workspace may not be less than 30 inches wide in front of the electric equipment. Distances shall be measured from the live parts if they are exposed, or from the enclosure front or opening if the live parts are enclosed. Concrete, brick, or tile walls are considered to be grounded. Working space is not required in back of assemblies such as dead-front switchboards or motor control centers where there are no renewable or adjustable parts such as fuses or switches on the back and where all connections are accessible from locations other than the back.

Table S - 1 — *Working Clearances*

Nominal voltage to ground	Minimum clear distance for condition (ft)		
	(a)	(b)	(c)
0 - 150................	[1]3	[1]3	3
151 - 600.......	[1]3	3 1/2	4

FOOTNOTE: [1]Minimum clear distances may be 2 feet 6 inches for installations built prior to April 16, 1981.

FOOTNOTE: [2]Conditions (a), (b), and (c), are as follows: (a) Exposed live parts on one side and no live or grounded parts on the other side of the working space, or exposed live parts on both sides effectively guarded by suitable wood or other insulating material. Insulated wire or insulated busbars operating at not over 300 volts are not considered live parts. (b) Exposed live parts on one side and grounded parts on the other side. (c) Exposed live parts on both sides of the workspace [not guarded as provided in Condition (a)] with the operator between.

(ii) *Clear spaces.* Working space required by this subpart may not be used for storage. When normally enclosed live parts are exposed for inspection or servicing, the working space, if in a passageway or general open space, shall be suitably guarded.

(iii) *Access and entrance to working space.* At least one entrance of sufficient area shall be provided to give access to the working space about electric equipment.

(iv) *Front working space.* Where there are live parts normally exposed on the front of switchboards or motor control centers, the working space in front of such equipment may not be less than 3 feet.

(v) *Illumination.* Illumination shall be provided for all working spaces about service equipment, switchboards, panelboards, and motor control centers installed indoors.

(vi) *Headroom.* The minimum headroom of working spaces about service equipment, switchboards, panel-boards, or motor control centers shall be 6 feet 3 inches.

Note: As used in this section a motor control center is an assembly of one or more enclosed sections having a common power bus and principally containing motor control units.

(2) *Guarding of live parts.* (i) Except as required or permitted elsewhere in this subpart, live parts of electric equipment operating at 50 volts or more shall be guarded against accidental contact by approved cabinets or other forms of approved enclosures, or by any of the following means:

(A) By location in a room, vault, or similar enclosure that is accessible only to qualified persons.

(B) By suitable permanent, substantial partitions or screens so arranged that only qualified persons will have access to the space within reach of the live parts. Any openings in such partitions or screens shall be so sized and located that persons are not likely to come into accidental contact with the live parts or to bring conducting objects into contact with them.

(C) By location on a suitable balcony, gallery, or platform so elevated and arranged as to exclude unqualified persons.

(D) By elevation of 8 feet or more above the floor or other working surface.

(ii) In locations where electric equipment would be exposed to physical damage, enclosures or guards shall be so arranged and of such strength as to prevent such damage.

(iii) Entrances to rooms and other guarded locations containing exposed live parts shall be marked with conspicuous warning signs forbidding unqualified persons to enter.

(h) *Over 600 volts, nominal* — (1) *General.* Conductors and equipment used on circuits exceeding 600 volts, nominal, shall comply with all applicable provisions of paragraphs (a) through (g) of this section and with the following provisions which supplement or modify those requirements. The provisions of paragraphs (h)(2), (h)(3), and (h)(4) of this section do not apply to equipment on the supply side of the service conductors.

(2) *Enclosure for electrical installations.* Electrical installations in a vault, room, closet or in an area surrounded by a wall, screen, or fence, access to which is controlled by lock and key or other approved means, are considered to be accessible to qualified persons only. A wall, screen, or fence less than 8 feet in height is not considered to prevent access unless it has other features that provide a degree of isolation equivalent to an 8 foot fence. The entrances to all buildings, rooms, or enclosures containing exposed live parts or exposed conductors operating at over 600 volts, nominal, shall be kept locked or shall be under the observation of a qualified person at all times.

(i) *Installations accessible to qualified persons only.* Electrical installations having exposed live parts shall be accessible to qualified persons only and shall comply with the applicable provisions of paragraph (h)(3) of this section.

(ii) *Installations accessible to unqualified persons.* Electrical installations that are open to unqualified persons shall be made with metal-enclosed equipment or shall be enclosed in a vault or in an area, access to which is controlled by a lock. If metal-enclosed equipment is installed so that the bottom of the enclosure is less than 8 feet above the floor, the door or cover shall be kept locked. Metal-enclosed switchgear, unit substations, transformers, pull boxes, connection boxes, and other similar associated equipment shall be marked with appropriate caution signs. If equipment is exposed to physical damage from vehicular traffic, suitable guards shall be provided to prevent such damage. Ventilating or similar openings in metal-enclosed equipment shall be designed so that foreign objects inserted through these openings will be deflected from energized parts.

(3) *Workspace about equipment.* Sufficient space shall be provided and maintained about electric equipment to permit ready and safe operation and maintenance of such equipment. Where energized parts are exposed, the minimum clear workspace may not be less than 6 feet 6 inches high (measured vertically from the floor or platform), or less than 3 feet wide (measured parallel to the equipment). The depth shall be as required in Table S - 2. The workspace shall be adequate to permit at least a 90-degree opening of doors or hinged panels.

(i) *Working space.* The minimum clear working space in front of electric equipment such as switchboards, control panels, switches, circuit breakers, motor controllers, relays, and similar equipment may not be less than specified in Table S - 2 unless otherwise specified in this subpart. Distances shall be measured from the live parts if they are exposed, or from the enclosure front or opening if the live parts are enclosed. However, working space is not required in back of equipment such as deadfront switchboards or control assemblies where there are no renewable or adjustable parts (such as fuses or switches) on the back and where all connections are accessible from locations other than the back. Where rear access is required to work on de-energized parts on the back of enclosed equipment, a minimum working space of 30 inches horizontally shall be provided.

Table S - 2 — *Minimum Depth of Clear Working Space in Front of Electric Equipment*

Nominal voltage to ground	Conditions (ft)		
	(a)	(b)	(c)
601 to 2,500............	3	4	5
2,501 to 9,000...........	4	5	6
9,001 to 25,000.........	5	6	9
25,001 to 75kV[1].........	6	8	10
Above 75kV[1]..............	8	10	12

FOOTNOTE: [1]Minimum depth of clear working space in front of electric equipment with a nominal voltage to ground above 25,000 volts may be the same as for 25,000 volts under Conditions (a), (b), and (c) for installations built prior to April 16, 1981.

FOOTNOTE: ²Conditions (a), (b), and (c) are as follows: (a) Exposed live parts on one side and no live or grounded parts on the other side of the working space, or exposed live parts on both sides effectively guarded by suitable wood or other insulating materials. Insulated wire or insulated busbars operating at not over 300 volts are not considered live parts. (b) Exposed live parts on one side and grounded parts on the other side. Concrete, brick, or tile walls will be considered as grounded surfaces. (c) Exposed live parts on both sides of the workspace not guarded as provided in Condition (a) with the operator between.

(ii) *Illumination.* Adequate illumination shall be provided for all working spaces about electric equipment. The lighting outlets shall be so arranged that persons changing lamps or making repairs on the lighting system will not be endangered by live parts or other equipment. The points of control shall be so located that persons are not likely to come in contact with any live part or moving part of the equipment while turning on the lights.

(iii) *Elevation of unguarded live parts.* Unguarded live parts above working space shall be maintained at elevations not less than specified in Table S - 3.

Table S - 3 — *Elevation of Unguarded Energized Parts Above Working Space*

Nominal voltage between phases	Minimum elevation
601 to 7,500.................	˙8 feet 6 inches.
7,501 to 35,000..............	9 feet.
Over 35kV....................	9 feet + 0.37 inches per kV above 35kV.

FOOTNOTE: ˙Note. — Minimum elevation may be 8 feet 0 inches for installations built prior to April 16, 1981 if the nominal voltage between phases is in the range of 601 - 6600 volts.

(4) *Entrance and access to workspace.* (See 1910.302(b)(3).)

(i) At least one entrance not less than 24 inches wide and 6 feet 6 inches high shall be provided to give access to the working space about electric equipment. On switchboard and control panels exceeding 48 inches in width, there shall be one entrance at each end of such board where practicable. Where bare energized parts at any voltage or insulated energized parts above 600 volts are located adjacent to such entrance, they shall be suitably guarded.

(ii) Permanent ladders or stairways shall be provided to give safe access to the working space around electric equipment installed on platforms, balconies, mezzanine floors, or in attic or roof rooms or spaces.

[46 FR 4056, Jan. 16, 1981; 46 FR 40185, Aug. 7, 1981]

Sec. 1910.304 Wiring design and protection.

(a) *Use and identification of grounded and grounding conductors.* (1) *Identification of conductors.* A conductor used as a grounded conductor shall be identifiable and distinguishable from all other conductors. A conductor used as an equipment grounding conductor shall be identifiable and distinguishable from all other conductors.

(2) *Polarity of connections.* No grounded conductor may be attached to any terminal or lead so as to reverse designated polarity.

(3) *Use of grounding terminals and devices.* A grounding terminal or grounding-type device on a receptacle, cord connector, or attachment plug may not be used for purposes other than grounding.

(b) *Branch Circuits* — (1) (See 29 CFR 1926 Subpart K, Electrical Standards for Construction)

(2) *Outlet devices.* Outlet devices shall have an ampere rating not less than the load to be served.

(c) *Outside conductors, 600 volts nominal, or less.* Paragraphs (c)(1), (c)(2), (c)(3), and (c)(4) of this section apply to branch circuit, feeder, and service conductors rated 600 volts, nominal, or less and run outdoors as open conductors. Paragraph (c)(5) applies to lamps installed under such conductors.

(1) *Conductors on poles.* Conductors supported on poles shall provide a horizontal climbing space not less than the following:

(i)   Power conductors below communication conductors — 30 inches.

(ii)   Power conductors alone or above communication conductors: 300 volts or less — 24 inches; more than 300 volts — 30 inches.

(iii) Communication conductors below power conductors with power conductors 300 volts or less — 24 inches; more than 300 volts — 30 inches.

(2) *Clearance from ground.* Open conductors shall conform to the following minimum clearances:

(i)   10 feet — above finished grade, sidewalks, or from any platform or projection from which they might be reached.

(ii)   12 feet — over areas subject to vehicular traffic other than truck traffic.

(iii) 15 feet — over areas other than those specified in paragraph (c)(2)(iv) of this section that are subject to truck traffic.

(iv)   18 feet — over public streets, alleys, roads, and driveways.

(3) *Clearance from building openings.* Conductors shall have a clearance of at least 3 feet from windows, doors, porches, fire escapes, or similar locations. Conductors run above the top level of a window are considered to be out of reach from that window and, therefore, do not have to be 3 feet away.

(4) *Clearance over roofs.* Conductors shall have a clearance of not less than 8 feet from the highest point of roofs over which they pass, except that:

(i) Where the voltage between conductors is 300 volts or less and the roof has a slope of not less than 4 inches in 12, the clearance from roofs shall be at least 3 feet, or

(ii) Where the voltage between conductors is 300 volts or less and the conductors do not pass over more than 4 feet of the overhang portion of the roof and they are terminated at a through-the-roof raceway or approved support, the clearance from roofs shall be at least 18 inches.

(5) *Location of outdoor lamps.* Lamps for outdoor lighting shall be located below all live conductors, transformers, or other electric equipment, unless such equipment is controlled by a disconnecting means that can be locked in the open position or unless adequate clearances or other safeguards are provided for relamping operations.

(d) *Services* — (1) *Disconnecting means* — (i) *General.* Means shall be provided to disconnect all conductors in a building or other structure from the service-entrance conductors. The disconnecting means shall plainly indicate whether it is in the open or closed position and shall be installed at a readily accessible location nearest the point of entrance of the service-entrance conductors.

(ii) *Simultaneous opening of poles.* Each service disconnecting means shall simultaneously disconnect all ungrounded conductors.

(2) *Services over 600 volts, nominal.* The following additional requirements apply to services over 600 volts, nominal.

(i) *Guarding.* Service-entrance conductors installed as open wires shall be guarded to make them accessible only to qualified persons.

(ii) *Warning signs.* Signs warning of high voltage shall be posted where other than qualified employees might come in contact with live parts.

(e) *Overcurrent protection.* (1) *600 volts, nominal, or less.* The following requirements apply to overcurrent protection of circuits rated 600 volts, nominal, or less.

(i) *Protection of conductors and equipment.* Conductors and equipment shall be protected from overcurrent in accordance with their ability to safely conduct current.

(ii) *Grounded conductors.* Except for motor running overload protection, overcurrent devices may not interrupt the continuity of the grounded conductor unless all conductors of the circuit are opened simultaneously.

(iii) *Disconnection of fuses and thermal cutouts.* Except for service fuses, all cartridge fuses which are accessible to other than qualified persons and all fuses and thermal cutouts on circuits over 150 volts to ground shall be provided with disconnecting means. This disconnecting means shall be installed so that the fuse or thermal cutout can be disconnected from its supply without disrupting service to equipment and circuits unrelated to those protected by the overcurrent device.

(iv) *Location in or on premises.* Overcurrent devices shall be readily accessible to each employee or authorized building management personnel. These overcurrent devices may not be located where they will be exposed to physical damage nor in the vicinity of easily ignitible material.

(v) *Arcing or suddenly moving parts.* Fuses and circuit breakers shall be so located or shielded that employees will not be burned or otherwise injured by their operation.

(vi) *Circuit breakers.* (A) Circuit breakers shall clearly indicate whether they are in the open (off) or closed (on) position.

(B) Where circuit breaker handles on switchboards are operated vertically rather than horizontally or rotationally, the up position of the handle shall be the closed (on) position. (See 1910.302(b)(3).)

(C) If used as switches in 120-volt, fluorescent lighting circuits, circuit breakers shall be approved for the purpose and marked "SWD." (See 1910.302(b)(3).)

(2) *Over 600 volts, nominal.* Feeders and branch circuits over 600 volts, nominal, shall have short-circuit protection.

(f) *Grounding.* Paragraphs (f)(1) through (f)(7) of this section contain grounding requirements for systems, circuits, and equipment.

(1) *Systems to be grounded.* The following systems which supply premises wiring shall be grounded:

(i) All 3-wire DC systems shall have their neutral conductor grounded.

(ii) Two-wire DC systems operating at over 50 volts through 300 volts between conductors shall be grounded unless:

(A) They supply only industrial equipment in limited areas and are equipped with a ground detector; or

(B) They are rectifier-derived from an AC system complying with paragraphs (f)(1)(iii), (f)(1)(iv), and (f)(1)(v) of this section; or

(C) They are fire-protective signaling circuits having a maximum current of 0.030 amperes.

(iii) AC circuits of less than 50 volts shall be grounded if they are installed as overhead conductors outside of buildings or if they are supplied by transformers and the transformer primary supply system is ungrounded or exceeds 150 volts to ground.

(iv) AC systems of 50 volts to 1000 volts shall be grounded under any of the following conditions, unless exempted by paragraph (f)(1)(v) of this section:

(A) If the system can be so grounded that the maximum voltage to ground on the ungrounded conductors does not exceed 150 volts;

(B) If the system is nominally rated 480Y/277 volt, 3-phase, 4-wire in which the neutral is used as a circuit conductor;

(C) If the system is nominally rated 240/120 volt, 3-phase, 4-wire in which the midpoint of one phase is used as a circuit conductor; or

(D) If a service conductor is uninsulated.

(v) AC systems of 50 volts to 1000 volts are not required to be grounded under any of the following conditions:

(A) If the system is used exclusively to supply industrial electric furnaces for melting, refining, tempering, and the like.

(B) If the system is separately derived and is used exclusively for rectifiers supplying only adjustable speed industrial drives.

(C) If the system is separately derived and is supplied by a transformer that has a primary voltage rating less than 1000 volts, provided all of the following conditions are met:

(1) The system is used exclusively for control circuits,

(2) The conditions of maintenance and supervision assure that only qualified persons will service the installation,

(3) Continuity of control power is required, and

(4) Ground detectors are installed on the control system.

(D) If the system is an isolated power system that supplies circuits in health care facilities.

(2) *Conductors to be grounded.* For AC premises wiring systems the identified conductor shall be grounded.

(3) *Grounding connections.* (i) For a grounded system, a grounding electrode conductor shall be used to connect both the equipment grounding conductor and the grounded circuit conductor to the grounding electrode. Both the equipment grounding conductor and the grounding electrode conductor shall be connected to the grounded circuit conductor on the supply side of the service disconnecting means, or on the supply side of the system disconnecting means or overcurrent devices if the system is separately derived.

(ii) For an ungrounded service-supplied system, the equipment grounding conductor shall be connected to the grounding electrode conductor at the service equipment. For an ungrounded separately derived system, the equipment grounding conductor shall be connected to the grounding electrode conductor at, or ahead of, the system disconnecting means or overcurrent devices.

(iii) On extensions of existing branch circuits which do not have an equipment grounding conductor, grounding-type receptacles may be grounded to a grounded cold water pipe near the equipment.

(4) *Grounding path.* The path to ground from circuits, equipment, and enclosures shall be permanent and continuous.

(5) *Supports, enclosures, and equipment to be grounded* — (i) *Supports and enclosures for conductors.* Metal cable trays, metal raceways, and metal enclosures for conductors shall be grounded, except that:

(A) Metal enclosures such as sleeves that are used to protect cable assemblies from physical damage need not be grounded; or

(B) Metal enclosures for conductors added to existing installations of open wire, knob-and-tube wiring, and nonmetallic-sheathed cable need not be grounded if all of the following conditions are met: (1) Runs are less than 25 feet; (2) enclosures are free from probable contact with ground, grounded metal, metal laths, or other conductive materials; and (3) enclosures are guarded against employee contact.

(ii) *Service equipment enclosures.* Metal enclosures for service equipment shall be grounded.

(iii) *Frames of ranges and clothes dryers.* Frames of electric ranges, wall-mounted ovens, counter-mounted cooking units, clothes dryers, and metal outlet or junction boxes which are part of the circuit for these appliances shall be grounded.

(iv) *Fixed equipment.* Exposed non-current-carrying metal parts of fixed equipment which may become energized shall be grounded under any of the following conditions:

(A) If within 8 feet vertically or 5 feet horizontally of ground or grounded metal objects and subject to employee contact.

(B) If located in a wet or damp location and not isolated.

(C) If in electrical contact with metal.

(D) If in a hazardous (classified) location.

(E) If supplied by a metal-clad, metal-sheathed, or grounded metal raceway wiring method.

(F) If equipment operates with any terminal at over 150 volts to ground; however, the following need not be grounded:

(1) Enclosures for switches or circuit breakers used for other than service equipment and accessible to qualified persons only;

(2) Metal frames of electrically heated appliances which are permanently and effectively insulated from ground; and

(3) The cases of distribution apparatus such as transformers and

capacitors mounted on wooden poles at a height exceeding 8 feet above ground or grade level.

(v) Equipment connected by cord and plug. Under any of the conditions described in paragraphs (f)(5)(v)(A) through (f)(5)(v)(C) of this section, exposed non-current-carrying metal parts of cord - and plug-connected equipment which may become energized shall be grounded.

(A) If in hazardous (classified) locations (see 1910.307).

(B) If operated at over 150 volts to ground, except for guarded motors and metal frames of electrically heated appliances if the appliance frames are permanently and effectively insulated from ground.

(C) If the equipment is of the following types:

(1) Refrigerators, freezers, and air conditioners;

(2) Clothes-washing, clothes-drying and dishwashing machines, sump pumps, and electrical aquarium equipment;

(3) Hand-held motor-operated tools;

(4) Motor-operated appliances of the following types: hedge clippers, lawn mowers, snow blowers, and wet scrubbers;

(5) Cord- and plug-connected appliances used in damp or wet locations or by employees standing on the ground or on metal floors or working inside of metal tanks or boilers;

(6) Portable and mobile X-ray and associated equipment;

(7) Tools likely to be used in wet and conductive locations; and

(8) Portable hand lamps.

Tools likely to be used in wet and conductive locations need not be grounded if supplied through an isolating transformer with an ungrounded secondary of not over 50 volts. Listed or labeled portable tools and appliances protected by an approved system of double insulation, or its equivalent, need not be grounded. If such a system is employed, the equipment shall be distinctively marked to indicate that the tool or appliance utilizes an approved system of double insulation.

(vi) *Nonelectrical equipment.* The metal parts of the following nonelectrical equipment shall be grounded: frames and tracks of electrically operated cranes; frames of nonelectrically driven elevator cars to which electric conductors are attached; hand operated metal shifting ropes or cables of electric elevators, and metal partitions, grill work, and similar metal enclosures around equipment of over 750 volts between conductors.

(6) *Methods of grounding fixed equipment.* (i) Non-current-carrying metal parts of fixed equipment, if required to be grounded by this subpart, shall be grounded by an equipment grounding conductor which is contained within the same raceway, cable, or cord, or runs with

119

or encloses the circuit conductors. For DC circuits only, the equipment grounding conductor may be run separately from the circuit conductors. (ii) Electric equipment is considered to be effectively grounded if it is secured to, and in electrical contact with, a metal rack or structure that is provided for its support and the metal rack or structure is grounded by the method specified for the non-current-carrying metal parts of fixed equipment in paragraph (f)(6)(i) of this section. For installations made before April 16, 1981, only, electric equipment is also considered to be effectively grounded if it is secured to, and in metallic contact with, the grounded structural metal frame of a building. Metal car frames supported by metal hoisting cables attached to or running over metal sheaves or drums of grounded elevator machines are also considered to be effectively grounded.

(7) *Grounding of systems and circuits of 1000 volts and over (high voltage.)* - (i) *General.* If high voltage systems are grounded, they shall comply with all applicable provisions of paragraphs (f)(1) through (f)(6) of this section as supplemented and modified by this paragraph (f)(7). (ii) Grounding of systems supplying portable or mobile equipment. (See 1910.302(b)(3).) Systems supplying portable or mobile high voltage equipment, other than substations installed on a temporary basis, shall comply with the following:

(A) Portable and mobile high voltage equipment shall be supplied from a system having its neutral grounded through an impedance. If a delta-connected high voltage system is used to supply the equipment, a system neutral shall be derived.

(B) Exposed non-current-carrying metal parts of portable and mobile equipment shall be connected by an equipment grounding conductor to the point at which the system neutral impedance is grounded.

(C) Ground-fault detection and relaying shall be provided to automatically deenergize any high voltage system component which has developed a ground fault. The continuity of the equipment grounding conductor shall be continuously monitored so as to deenergize automatically the high voltage feeder to the portable equipment upon loss of continuity of the equipment grounding conductor.

(D) The grounding electrode to which the portable or mobile equipment system neutral impedance is connected shall be isolated from and separated in the ground by at least 20 feet from any other system or equipment grounding electrode, and there shall be no direct connection between the grounding electrodes, such as buried pipe, fence, etc.

(iii) *Grounding of equipment.* All non-current-carrying metal parts of portable equipment and fixed equipment including their associated fences, housings, enclosures, and supporting structures shall be grounded. However, equipment which is guarded by location and isolated from ground need not be grounded. Additionally, pole-mounted distribution apparatus at a height exceeding 8 feet above ground or grade level need not be grounded.

[46 FR 4056, Jan. 16, 1981; 46 FR 40185, Aug. 7, 1981]
* [55 FR 32015, Aug. 6, 1990]

Sec. 1910.305 Wiring methods, components, and equipment for general use.

(a) *Wiring methods.* The provisions of this section do not apply to the conductors that are an integral part of factory-assembled equipment.

(1) *General requirements* - (i) Electrical continuity of metal raceways and enclosures. Metal raceways, cable armor, and other metal enclosures for conductors shall be metallically joined together into a continuous electric conductor and shall be so connected to all boxes, fittings, and cabinets as to provide effective electrical continuity.

(ii) *Wiring in ducts.* No wiring systems of any type shall be installed in ducts used to transport dust, loose stock or flammable vapors. No wiring system of any type may be installed in any duct used for vapor removal or for ventilation of commercial-type cooking equipment, or in any shaft containing only such ducts.

(2) *Temporary wiring.* Temporary electrical power and lighting wiring methods may be of a class less than would be required for a permanent installation. Except as specifically modified in this paragraph, all other requirements of this subpart for permanent wiring shall apply to temporary wiring installations.

(i) *Uses permitted, 600 volts, nominal, or less.* Temporary electrical power and lighting installations 600 volts, nominal, or less may be used only:

(A) During and for remodeling, maintenance, repair, or demolition of buildings, structures, or equipment, and similar activities;

(B) For experimental or development work, and

(C) For a period not to exceed 90 days for Christmas decorative lighting, carnivals, and similar purposes.

(ii) *Uses permitted, over 600 volts, nominal.* Temporary wiring over 600 volts, nominal, may be used only during periods of tests, experiments, or emergencies.

(iii) *General requirements for temporary wiring.*

(A) Feeders shall originate in an approved distribution center. The conductors shall be run as multiconductor cord or cable assemblies, or, where not subject to physical damage, they may be run as open conductors on insulators not more than 10 feet apart.

(B) *Branch circuits shall originate in an approved power outlet or panelboard.* Conductors shall be multiconductor cord or cable assemblies or open conductors. If run as open conductors they shall be fastened at ceiling height every 10 feet. No branch-circuit conductor may be laid on the floor. Each branch circuit that supplies receptacles or fixed equipment shall contain a separate equipment grounding conductor if run as open conductors.

(C) *Receptacles shall be of the grounding type.* Unless installed in a complete metallic raceway, each branch circuit shall contain a separate equipment grounding conductor and all receptacles shall be electrically connected to the grounding conductor.

(D) No bare conductors nor earth returns may be used for the wiring of any temporary circuit.

(E) Suitable disconnecting switches or plug connectors shall be installed to permit the disconnection of all ungrounded conductors of each temporary circuit.

(F) *Lamps for general illumination shall be protected from accidental contact or breakage.* Protection shall be provided by elevation of at least 7 feet from normal working surface or by a suitable fixture or lampholder with a guard.

(G) Flexible cords and cables shall be protected from accidental damage. Sharp corners and projections shall be avoided. Where passing through doorways or other pinch points, flexible cords and cables shall be provided with protection to avoid damage.

(3) *Cable trays.* (i) *Uses permitted.* (a) Only the following may be installed in cable tray systems:

(1) Mineral-insulated metal-sheathed cable (Type MI);

(2) Armored cable (Type AC);

(3) Metal-clad cable (Type MC);

(4) Power-limited tray cable (Type PLTC);

(5) Nonmetallic-sheathed cable (Type NM or NMC);

(6) Shielded nonmetallic-sheathed cable (Type SNM);

(7) Multiconductor service-entrance cable (Type SE or USE);

(8) Multiconductor underground feeder and branch-circuit cable (Type UF);

(9) Power and control tray cable (Type TC);

(10) Other factory-assembled, multiconductor control, signal, or

power cables which are specifically approved for installation in cable trays; or

(11) Any approved conduit or raceway with its contained conductors.

(b) In industrial establishments only, where conditions of maintenance and supervision assure that only qualified persons will service the installed cable tray system, the following cables may also be installed in ladder, ventilated trough, or 4 inch ventilated channel-type cable trays:

(1) Single conductor cables which are 250 MCM or larger and are Types RHH, RHW, MV, USE, or THW, and other 250 MCM or larger single conductor cables if specifically approved for installation in cable trays. Where exposed to direct rays of the sun, cables shall be sunlight-resistant.

(2) Type MV cables, where exposed to direct rays of the sun, shall be sunlight-resistant.

(c) Cable trays in hazardous (classified) locations shall contain only the cable types permitted in such locations.

(ii) Uses not permitted. Cable tray systems may not be used in hoistways or where subjected to severe physical damage.

(4) *Open wiring on insulators* — (i) *Uses permitted.* Open wiring on insulators is only permitted on systems of 600 volts, nominal, or less for industrial or agricultural establishments and for services.

(ii) *Conductor supports.* Conductors shall be rigidly supported on noncombustible, nonabsorbent insulating materials and may not contact any other objects.

(iii) *Flexible nonmetallic tubing.* In dry locations where not exposed to severe physical damage, conductors may be separately enclosed in flexible nonmetallic tubing. The tubing shall be in continuous lengths not exceeding 15 feet and secured to the surface by straps at intervals not exceeding 4 feet 6 inches.

(iv) *Through walls, floors, wood cross members, etc.* Open conductors shall be separated from contact with walls, floors, wood cross members, or partitions through which they pass by tubes or bushings of noncombustible, nonabsorbent insulating material. If the bushing is shorter than the hole, a waterproof sleeve of nonconductive material shall be inserted in the hole and an insulating bushing slipped into the sleeve at each end in such a manner as to keep the conductors absolutely out of contact with the sleeve. Each conductor shall be carried through a separate tube or sleeve.

(v) *Protection from physical damage.* Conductors within 7 feet from the floor are considered exposed to physical damage. Where open conductors cross ceiling joints and wall studs and are exposed to physical damage, they shall be protected.

(b) *Cabinets, boxes, and fittings* — (1) *Conductors entering boxes, cabinets, or fittings.* Conductors entering boxes, cabinets, or fittings shall also be protected from abrasion, and openings through which conductors enter shall be effectively closed. Unused openings in cabinets, boxes, and fittings shall be effectively closed.

(2) *Covers and canopies.* All pull boxes, junction boxes, and fittings shall be provided with covers approved for the purpose. If metal covers are used they shall be grounded. In completed installations each outlet box shall have a cover, faceplate, or fixture canopy. Covers of outlet boxes having holes through which flexible cord pendants pass shall be provided with bushings designed for the purpose or shall have smooth, well-rounded surfaces on which the cords may bear.

(3) *Pull and junction boxes for systems over 600 volts, nominal.* In addition to other requirements in this section for pull and junction boxes, the following shall apply to these boxes for systems over 600 volts, nominal:

(i) Boxes shall provide a complete enclosure for the contained conductors or cables.

(ii) Boxes shall be closed by suitable covers securely fastened in place. Underground box covers that weigh over 100 pounds meet this requirement. Covers for boxes shall be permanently marked "HIGH VOLTAGE." The marking shall be on the outside of the box cover and shall be readily visible and legible.

(c) *Switches* — (1) *Knife switches.* Single-throw knife switches shall be so connected that the blades are dead when the switch is in the open position. Single-throw knife switches shall be so placed that gravity will not tend to close them. Single-throw knife switches approved for use in the inverted position shall be provided with a locking device that will ensure that the blades remain in the open position when so set. Double-throw knife switches may be mounted so that the throw will be either vertical or horizontal. However, if the throw is vertical a locking device shall be provided to ensure that the blades remain in the open position when so set.

(2) *Faceplates for flush-mounted snap switches.* Flush snap switches that are mounted in ungrounded metal boxes and located within reach of conducting floors or other conducting surfaces shall be provided with faceplates of nonconducting, noncombustible material.

(d) *Switchboards and panelboards.* Switchboards that have any exposed live parts shall be located in permanently dry locations and accessible only to qualified persons. Panelboards shall be mounted in cabinets, cutout boxes, or enclosures approved for the purpose and shall be dead front. However, panelboards other than the dead front externally-operable type are permitted where accessible only to qualified persons. Exposed blades of knife switches shall be dead when open.

(e) *Enclosures for damp or wet locations.* (1) Cabinets, cutout boxes, fittings, boxes, and panelboard enclosures in damp or wet locations shall be installed so as to prevent moisture or water from entering and accumulating within the enclosures. In wet locations the enclosures shall be weatherproof.

(2) Switches, circuit breakers, and switchboards installed in wet locations shall be enclosed in weatherproof enclosures.

(f) *Conductors for general wiring.* All conductors used for general wiring shall be insulated unless otherwise permitted in this Subpart. The conductor insulation shall be of a type that is approved for the voltage, operating temperature, and location of use. Insulated conductors shall be distinguishable by appropriate color or other suitable means as being grounded conductors, ungrounded conductors, or equipment grounding conductors.

(g) *Flexible cords and cables* — Use of flexible cords and cables. (i) Flexible cords and cables shall be approved and suitable for conditions of use and location. Flexible cords and cables shall be used only for:

(A) Pendants;

(B) Wiring of fixtures;

(C) Connection of portable lamps or appliances;

(D) Elevator cables;

(E) Wiring of cranes and hoists;

(F) Connection of stationary equipment to facilitate their frequent interchange;

(G) Prevention of the transmission of noise or vibration;

(H) Appliances where the fastening means and mechanical connections are designed to permit removal for maintenance and repair; or

(i)   Data processing cables approved as a part of the data processing system.

(ii) If used as permitted in paragraphs (g)(1)(i)(c), (g)(1)(i)(f), or (g)(1)(i)(h) of this section, the flexible cord shall be equipped with an attachment plug and shall be energized from an approved receptacle outlet.

(iii) Unless specifically permitted in paragraph (g)(1)(i) of this section, flexible cords and cables may not be used:

(A) As a substitute for the fixed wiring of a structure;

(B) Where run through holes in walls, ceilings, or floors;

(C) Where run through doorways, windows, or similar openings;

(D) Where attached to building surfaces; or

(E) Where concealed behind building walls, ceilings, or floors.

(iv) Flexible cords used in show windows and showcases shall be Type S, SO, SJ, SJO, ST, STO, SJT, SJTO, or AFS except for the wiring of chain-supported lighting fixtures and supply cords for portable lamps and other merchandise being displayed or exhibited.

(2) *Identification, splices, and terminations.* (i) A conductor of a flexible cord or cable that is used as a grounded conductor or an equipment grounding conductor shall be distinguishable from other conductors. Types SJ, SJO, SJT, SJTO, S, SO, ST, and STO shall be durably marked on the surface with the type designation, size, and number of conductors.

(ii) Flexible cords shall be used only in continuous lengths without splice or tap. Hard service flexible cords No. 12 or larger may be repaired if spliced so that the splice retains the insulation, outer sheath properties, and usage characteristics of the cord being spliced.

(iii) Flexible cords shall be connected to devices and fittings so that strain relief is provided which will prevent pull from being directly transmitted to joints or terminal screws.

(h) *Portable cables over 600 volts, nominal.* Multiconductor portable cable for use in supplying power to portable or mobile equipment at over 600 volts, nominal, shall consist of No. 8 or larger conductors employing flexible stranding. Cables operated at over 2,000 volts shall be shielded for the purpose of confining the voltage stresses to the insulation. Grounding conductors shall be provided. Connectors for these cables shall be of a locking type with provisions to prevent their opening or closing while energized. Strain relief shall be provided at connections and terminations. Portable cables may not be operated with splices unless the splices are of the permanent molded, vulcanized, or other approved type. Termination enclosures shall be suitably marked with a high voltage hazard warning, and terminations shall be accessible only to authorized and qualified personnel.

(i) *Fixture wires* — (1) *General.* Fixture wires shall be approved for the voltage, temperature, and location of use. A fixture wire which is used as a grounded conductor shall be identified.

(2) *Uses permitted.* Fixture wires may be used:

(i) For installation in lighting fixtures and in similar equipment where enclosed or protected and not subject to bending or twisting in use; or

(ii) For connecting lighting fixtures to the branch-circuit conductors supplying the fixtures.

(3) *Uses not permitted.* Fixture wires may not be used as branch-circuit conductors except as permitted for Class 1 power limited circuits.

(j) *Equipment for general use* — (1) *Lighting fixtures, lampholders, lamps, and receptacles.* (i) Fixtures, lampholders, lamps, rosettes, and receptacles may have no live parts normally exposed to employee contact. However, rosettes and cleat-type lampholders and receptacles located at least 8 feet above the floor may have exposed parts.

(ii) Handlamps of the portable type supplied through flexible cords shall be equipped with a handle of molded composition or other material approved for the purpose, and a substantial guard shall be attached to the lampholder or the handle.

(iii) Lampholders of the screw-shell type shall be installed for use as lampholders only. Lampholders installed in wet or damp locations shall be of the weatherproof type.

(iv) Fixtures installed in wet or damp locations shall be approved for the purpose and shall be so constructed or installed that water cannot enter or accumulate in wireways, lampholders, or other electrical parts.

(2) *Receptacles, cord connectors, and attachment plugs (caps).* (i) Receptacles, cord connectors, and attachment plugs shall be constructed so that no receptacle or cord connector will accept an attachment plug with a different voltage or current rating than that for which the device is intended. However, a 20-ampere T-slot receptacle or cord connector may accept a 15-ampere attachment plug of the same voltage rating.

(ii) A receptacle installed in a wet or damp location shall be suitable for the location.

(3) *Appliances.* (i) Appliances, other than those in which the current-carrying parts at high temperatures are necessarily exposed, may have no live parts normally exposed to employee contact.

(ii) A means shall be provided to disconnect each appliance.

(iii) Each appliance shall be marked with its rating in volts and amperes or volts and watts.

(4) *Motors.* This paragraph applies to motors, motor circuits, and controllers.

(i) *In sight from.* If specified that one piece of equipment shall be "in sight from" another piece of equipment, one shall be visible and not more than 50 feet from the other.

(ii) *Disconnecting means.* (A) A disconnecting means shall be located in sight from the controller location. However, a single disconnecting means may be located adjacent to a group of coordinated controllers mounted adjacent to each other on a multi-motor continuous process machine. The controller disconnecting means for motor branch circuits over 600 volts, nominal, may be out of sight of the controller, if the controller is marked with a warning label giving the location and identification of the disconnecting means which is to be locked in the open position.

(B) The disconnecting means shall disconnect the motor and the controller from all ungrounded supply conductors and shall be so designed that no pole can be operated independently.

(C) If a motor and the driven machinery are not in sight from the controller location, the installation shall comply with one of the following conditions:

(1) The controller disconnecting means shall be capable of being locked in the open position.

(2) A manually operable switch that will disconnect the motor from its source of supply shall be placed in sight from the motor location.

(D) The disconnecting means shall plainly indicate whether it is in the open (off) or closed (on) position.

(E) The disconnecting means shall be readily accessible. If more than one disconnect is provided for the same equipment, only one need be readily accessible.

(F) An individual disconnecting means shall be provided for each motor, but a single disconnecting means may be used for a group of motors under any one of the following conditions:

(1) If a number of motors drive special parts of a single machine or piece of apparatus, such as a metal or woodworking machine, crane, or hoist;

(2) If a group of motors is under the protection of one set of branch-circuit protective devices; or

(3) If a group of motors is in a single room in sight from the location of the disconnecting means.

(iii) Motor overload, short-circuit, and ground-fault protection. Motors, motor-control apparatus, and motor branch-circuit conductors shall be protected against overheating due to motor overloads or failure to start, and against short-circuits or ground faults. These provisions shall not require overload protection that will stop a motor where a shutdown is likely to introduce additional or increased hazards, as in the case of fire pumps, or where continued operation of a motor is necessary for a safe shutdown of equipment or process and motor overload sensing devices are connected to a supervised alarm.

(iv) *Protection of live parts — all voltages.* (A) Stationary motors having commutators, collectors, and brush rigging located inside of motor end brackets and not conductively connected to supply circuits operating at more than 150 volts to ground need not have such parts guarded. Exposed live parts of motors and controllers operating at 50 volts or more between terminals shall be guarded against accidental contact by any of the following:

(1) By installation in a room or enclosure that is accessible only to qualified persons;

(2) By installation on a suitable balcony, gallery, or platform, so elevated and arranged as to exclude unqualified persons; or

(3) By elevation 8 feet or more above the floor.

(B) Where live parts of motors or controllers operating at over 150 volts to ground are guarded against accidental contact only by location, and where adjustment or other attendance may be necessary during the operation of the apparatus, suitable insulating mats or platforms shall be provided so that the attendant cannot readily touch live parts unless standing on the mats or platforms.

(5) Transformers. (i) The following paragraphs cover the installation of all transformers except the following:

(A) Current transformers;

(B) Dry-type transformers installed as a component part of other apparatus;

(C) Transformers which are an integral part of an X-ray, high frequency, or electrostatic-coating apparatus;

(D) Transformers used with Class 2 and Class 3 circuits, sign and outline lighting, electric discharge lighting, and power-limited fire-protective signalling circuits; and

(E) Liquid-filled or dry-type transformers used for research, development, or testing, where effective safeguard arrangements are provided.

(ii) The operating voltage of exposed live parts of transformer installations shall be indicated by warning signs or visible markings on the equipment or structure.

(iii) Dry-type, high fire point liquid-insulated, and askarel-insulated transformers installed indoors and rated over 35kV shall be in a vault.

(iv) If they present a fire hazard to employees, oil-insulated transformers installed indoors shall be in a vault.

(v) Combustible material, combustible buildings and parts of buildings, fire escapes, and door and window openings shall be safeguarded from fires which may originate in oil-insulated transformers attached to or adjacent to a building or combustible material.

(vi) Transformer vaults shall be constructed so as to contain fire and combustible liquids within the vault and to prevent unauthorized access. Locks and latches shall be so arranged that a vault door can be readily opened from the inside.

(vii) Any pipe or duct system foreign to the vault installation may not enter or pass through a transformer vault.

(viii) Materials may not be stored in transformer vaults.

(6) *Capacitors.* (i) All capacitors, except surge capacitors or capacitors included as a component part of other apparatus, shall be provided with an automatic means of draining the stored charge after the capacitor is disconnected from its source of supply.

(ii) Capacitors rated over 600 volts, nominal, shall comply with the following additional requirements:

(A) Isolating or disconnecting switches (with no interrupting rating) shall be interlocked with the load interrupting device or shall be provided with prominently displayed caution signs to prevent switching load current.

(B) For series capacitors (see 1910.302(b)(3)), the proper switching shall be assured by use of at least one of the following:

(1) Mechanically sequenced isolating and bypass switches,

(2) Interlocks, or

(3) Switching procedure prominently displayed at the switching location.

(7) *Storage batteries.* Provisions shall be made for sufficient diffusion and ventilation of gases from storage batteries to prevent the accumulation of explosive mixtures.

[46 FR 4056, Jan. 16, 1981; 46 FR 40185, Aug. 7, 1981]

Sec. 1910.306 Specific purpose equipment and installations.
(a) *Electric signs and outline lighting* —(1) *Disconnecting means.* Signs operated by electronic or electromechanical controllers located outside the sign shall have a disconnecting means located inside the controller enclosure or within sight of the controller location, and it shall be capable of being locked in the open position. Such disconnecting means shall have no pole that can be operated independently, and it shall open all ungrounded conductors that supply the controller and sign. All other signs, except the portable type, and all outline lighting installations shall have an externally operable disconnecting means

which can open all ungrounded conductors and is within the sight of the sign or outline lighting it controls.

(2) Doors or covers giving access to uninsulated parts of indoor signs or outline lighting exceeding 600 volts and accessible to other than qualified persons shall either be provided with interlock switches to disconnect the primary circuit or shall be so fastened that the use of other than ordinary tools will be necessary to open them.

(b) *Cranes and hoists.* This paragraph applies to the installation of electric equipment and wiring used in connection with cranes, monorail hoists, hoists, and all runways.

(1) *Disconnecting means.* A readily accessible disconnecting means (i) shall be provided between the runway contact conductors and the power supply.

(ii) Another disconnecting means, capable of being locked in the open position, shall be provided in the leads from the runway contact conductors or other power supply on any crane or monorail hoist.

(A) If this additional disconnecting means is not readily accessible from the crane or monorail hoist operating station, means shall be provided at the operating station to open the power circuit to all motors of the crane or monorail hoist.

(B) The additional disconnect may be omitted if a monorail hoist or hand-propelled crane bridge installation meets all of the following:

(1) The unit is floor controlled;

(2) The unit is within view of the power supply disconnecting means; and

(3) No fixed work platform has been provided for servicing the unit.

(2) *Control.* A limit switch or other device shall be provided to prevent the load block from passing the safe upper limit of travel of any hoisting mechanism.

(3) *Clearance.* The dimension of the working space in the direction of access to live parts which may require examination, adjustment, servicing, or maintenance while alive shall be a minimum of 2 feet 6 inches. Where controls are enclosed in cabinets, the door(s) shall either open at least 90 degrees or be removable.

(c) *Elevators, dumbwaiters, escalators, and moving walks* —

(1) *Disconnecting means.* Elevators, dumbwaiters, escalators, and moving walks shall have a single means for disconnecting all ungrounded main power supply conductors for each unit.

(2) *Warning signs.* If interconnections between control panels are necessary for operation of the system on a multicar installation that remains energized from a source other than the disconnecting means, a warning sign shall be mounted on or adjacent to the disconnecting

means. The sign shall be clearly legible and shall read "Warning — Parts of the control panel are not de-energized by this switch." (See 1910.302(b)(3).)

(3) *Control panels.* If control panels are not located in the same space as the drive machine, they shall be located in cabinets with doors or panels capable of being locked closed.

(d) *Electric welders — disconnecting means.* (1) A disconnecting means shall be provided in the supply circuit for each motor-generator arc welder, and for each AC transformer and DC rectifier arc welder which is not equipped with a disconnect mounted as an intergral part of the welder.

(2) A switch or circuit breaker shall be provided by which each resistance welder and its control equipment can be isolated from the supply circuit. The ampere rating of this disconnecting means may not be less than the supply conductor ampacity.

(e) *Data processing systems — disconnecting means.* A disconnecting means shall be provided to disconnect the power to all electronic equipment in data processing or computer rooms. This disconnecting means shall be controlled from locations readily accessible to the operator at the principal exit doors. There shall also be a similar disconnecting means to disconnect the air conditioning system serving this area.

(f) *X-Ray equipment.* This paragraph applies to X-ray equipment for other than medical or dental use.

(1) *Disconnecting means.* (i) *A disconnecting means shall be provided in the supply circuit.* The disconnecting means shall be operable from a location readily accessible from the X-ray control. For equipment connected to a 120-volt branch circuit of 30 amperes or less, a grounding-type attachment plug cap and receptacle of proper rating may serve as a disconnecting means.

(ii) If more than one piece of equipment is operated from the same high-voltage circuit, each piece or each group of equipment as a unit shall be provided with a high-voltage switch or equivalent disconnecting means. This disconnecting means shall be constructed, enclosed, or located so as to avoid contact by employees with its live parts.

(2) *Control* — (i) *Radiographic and fluoroscopic types.* Radiographic and fluoroscopic-type equipment shall be effectively enclosed or shall have interlocks that de-energize the equipment automatically to prevent ready access to live current-carrying parts.

(ii) *Diffraction and irradiation types.* Diffraction- and irradiation-type equipment shall be provided with a means to indicate when it is energized unless the equipment or installation is effectively enclosed or is provided with interlocks to prevent access to live current-carrying parts during operation.

(g) Induction and dielectric heating equipment — (1) *Scope.* Paragraphs (g)(2) and (g)(3) of this section cover induction and dielectric heating equipment and accessories for industrial and scientific applications, but not for medical or dental applications or for appliances.

(2) *Guarding and grounding.* (i) *Enclosures.* The converting apparatus (including the DC line) and high-frequency electric circuits (excluding the output circuits and remote-control circuits) shall be completely contained within enclosures of noncombustible material.

(ii) *Panel controls.* All panel controls shall be of dead-front construction.

(iii) *Access to internal equipment.* Where doors are used for access to voltages from 500 to 1000 volts AC or DC, either door locks or interlocks shall be provided. Where doors are used for access to voltages of over 1000 volts AC or DC, either mechanical lockouts with a disconnecting means to prevent access until voltage is removed from the cubicle, or both door interlocking and mechanical door locks, shall be provided.

(iv) *Warning labels.* "Danger" labels shall be attached on the equipment and shall be plainly visible even when doors are open or panels are removed from compartments containing voltages of over 250 volts AC or DC.

(v) *Work applicator shielding.* Protective cages or adequate shielding shall be used to guard work applicators other than induction heating coils. Induction heating coils shall be protected by insulation and/or refractory materials. Interlock switches shall be used on all hinged access doors, sliding panels, or other such means of access to the applicator. Interlock switches shall be connected in such a manner as to remove all power from the applicator when any one of the access doors or panels is open. Interlocks on access doors or panels are not required if the applicator is an induction heating coil at DC ground potential or operating at less than 150 volts AC.

(vi) *Disconnecting means.* A readily accessible disconnecting means shall be provided by which each unit of heating equipment can be isolated from its supply circuit.

(3) *Remote control.* If remote controls are used for applying power, a selector switch shall be provided and interlocked to provide power from only one control point at a time. Switches operated by foot pressure shall be provided with a shield over the contact button to avoid accidental closing of the switch.

(h) *Electrolytic cells.* (1) *Scope.* These provisions for electrolytic cells apply to the installation of the electrical components and accessory equipment of electrolytic cells, electrolytic cell lines, and process power supply for the production of aluminum, cadmium, chlorine, copper, fluorine, hydrogen peroxide, magnesium, sodium, sodium chlorate, and zinc. Cells used as a source of electric energy and for electroplating processes and cells used for production of hydrogen are not covered by these provisions.

(2) *Definitions applicable to this paragraph.*

Cell line: An assembly of electrically interconnected electrolytic cells supplied by a source of direct-current power.

Cell line attachments and auxiliary equipment: Cell line attachments and auxiliary equipment include, but are not limited to: auxiliary tanks; process piping; duct work; structural supports; exposed cell line conductors; conduits and other raceways; pumps; positioning equipment and cell cutout or by-pass electrical devices. Auxiliary equipment also includes tools, welding machines, crucibles, and other portable equipment used for operation and maintenance within the electrolytic cell line working zone. In the cell line working zone, auxiliary equipment includes the exposed conductive surfaces of ungrounded cranes and crane-mounted cell-servicing equipment.

Cell line working zone: The cell line working zone is the space envelope wherein operation or maintenance is normally performed on or in the vicinity of exposed energized surfaces of cell lines or their attachments.

Electrolytic Cells: A receptacle or vessel in which electrochemical reactions are caused by applying energy for the purpose of refining or producing usable materials.

(3) *Application.* Installations covered by paragraph (h) of this section shall comply with all applicable provisions of this subpart, except as follows:

(i) Overcurrent protection of electrolytic cell DC process power circuits need not comply with the requirements of 1910.304(e).

(ii) Equipment located or used within the cell line working zone or associated with the cell line DC power circuits need not comply with the provisions of 1910.304(f).

(iii) Electrolytic cells, cell line conductors, cell line attachments, and the wiring of auxiliary equipment and devices within the cell line working zone need not comply with the provisions of 1910.303, and 1910.304 (b) and (c).

(4) *Disconnecting means.* (i) If more than one DC cell line process

power supply serves the same cell line, a disconnecting means shall be provided on the cell line circuit side of each power supply to disconnect it from the cell line circuit.

(ii) Removable links or removable conductors may be used as the disconnecting means.

(5) *Portable electric equipment.* (i) The frames and enclosures of portable electric equipment used within the cell line working zone may not be grounded. However, these frames and enclosures may be grounded if the cell line circuit voltage does not exceed 200 volts DC or if the frames are guarded.

(ii) Ungrounded portable electric equipment shall be distinctively marked and may not be interchangeable with grounded portable electric equipment.

(6) *Power supply circuits and receptacles for portable electric equipment.* (i) Circuits supplying power to ungrounded receptacles for hand-held, cord- and plug-connected equipment shall be electrically isolated from any distribution system supplying areas other than the cell line working zone and shall be ungrounded. Power for these circuits shall be supplied through isolating transformers.

(ii) Receptacles and their mating plugs for ungrounded equipment may not have provision for a grounding conductor and shall be of a configuration which prevents their use for equipment required to be grounded.

(iii) Receptacles on circuits supplied by an isolating transformer with an ungrounded secondary shall have a distinctive configuration, shall be distinctively marked, and may not be used in any other location in the plant.

(7) *Fixed and portable electric equipment.* (i) AC systems supplying fixed and portable electric equipment within the cell line working zone need not be grounded.

(ii) Exposed conductive surfaces, such as electric equipment housings, cabinets, boxes, motors, raceways and the like that are within the cell line working zone need not be grounded.

(iii) Auxiliary electrical devices, such as motors, transducers, sensors, control devices, and alarms, mounted on an electrolytic cell or other energized surface, shall be connected by any of the following means:

(A) Multiconductor hard usage or extra hard usage flexible cord;

(B) Wire or cable in suitable raceways; or

(C) Exposed metal conduit, cable tray, armored cable, or similar metallic systems installed with insulating breaks such that they will not cause a potentially hazardous electrical condition.

(iv) Fixed electric equipment may be bonded to the energized conductive surfaces of the cell line, its attachments, or auxiliaries. If fixed electric equipment is mounted on an energized conductive surface, it shall be bonded to that surface.

(8) *Auxiliary nonelectric connections.* Auxiliary nonelectric connections, such as air hoses, water hoses, and the like, to an electrolytic cell, its attachments, or auxiliary equipment may not have continuous conductive reinforcing wire, armor, braids, and the like. Hoses shall be of a nonconductive material.

(9) *Cranes and hoists.* (i) The conductive surfaces of cranes and hoists that enter the cell line working zone need not be grounded. The portion of an overhead crane or hoist which contacts an energized electrolytic cell or energized attachments shall be insulated from ground.

(ii) Remote crane or hoist controls which may introduce hazardous electrical conditions into the cell line working zone shall employ one or more of the following systems:

(A) Insulated and ungrounded control circuit;

(B) Nonconductive rope operator;

(C) Pendant pushbutton with nonconductive supporting means and having nonconductive surfaces or ungrounded exposed conductive surfaces; or

(D) Radio.

(i) *Electrically driven or controlled irrigation machines.* (See Sec. 1910.302(b)(3).)

(1) *Lightning protection.* If an electrically driven or controlled irrigation machine has a stationary point, a driven ground rod shall be connected to the machine at the stationary point for lightning protection.

(2) *Disconnecting means.* The main disconnecting means for a center pivot irrigation machine shall be located at the point of connection of electrical power to the machine and shall be readily accessible and capable of being locked in the open position. A disconnecting means shall be provided for each motor and controller.

(j) *Swimming pools, fountains, and similar installations* — (1) *Scope.* Paragraphs (j)(2) through (j)(5) of this section apply to electric wiring for and equipment in or adjacent to all swimming, wading, therapeutic, and decorative pools and fountains, whether permanently installed or storable, and to metallic auxiliary equipment, such as pumps, filters, and similar equipment. Therapeutic pools in health care facilities are exempt from these provisions.

(2) *Lighting and receptacles* — (i) *Receptacles.* A single receptacle of the locking and grounding type that provides power for a permanently installed swimming pool recirculating pump motor may be located not less than 5 feet from the inside walls of a pool. All other receptacles on the property shall be located at least 10 feet from the inside walls of a pool. Receptacles which are located within 15 feet of the inside walls of the pool shall be protected by ground-fault circuit interrupters.

Note: In determining these dimensions, the distance to be measured is the shortest path the supply cord of an appliance connected to the receptacle would follow without piercing a floor, wall, or ceiling of a building or other effective permanent barrier.

(ii) *Lighting fixtures and lighting outlets.* (A) Unless they are 12 feet above the maximum water level, lighting fixtures and lighting outlets may not be installed over a pool or over the area extending 5 feet horizontally from the inside walls of a pool. However, a lighting fixture or lighting outlet which has been installed before April 16, 1981, may be located less than 5 feet measured horizontally from the inside walls of a pool if it is at least 5 feet above the surface of the maximum water level and shall be rigidly attached to the existing structure. It shall also be protected by a ground-fault circuit interrupter installed in the branch circuit supplying the fixture.

(B) Unless installed 5 feet above the maximum water level and rigidly attached to the structure adjacent to or enclosing the pool, lighting fixtures and lighting outlets installed in the area extending between 5 feet and 10 feet horizontally from the inside walls of a pool shall be protected by a ground-fault circuit interrupter.

(3) Cord- and plug-connected equipment. Flexible cords used with the following equipment may not exceed 3 feet in length and shall have a copper equipment grounding conductor with a grounding-type attachment plug.

(i) Cord- and plug-connected lighting fixtures installed within 16 feet of the water surface of permanently installed pools.

(ii) Other cord- and plug-connected, fixed or stationary equipment used with permanently installed pools.

(4) *Underwater equipment.* (i) A ground-fault circuit interrupter shall be installed in the branch circuit supplying underwater fixtures operating at more than 15 volts. Equipment installed underwater shall be approved for the purpose.

(ii) No underwater lighting fixtures may be installed for operation at over 150 volts between conductors.

(5) *Fountains.* All electric equipment operating at more than 15

volts, including power supply cords, used with fountains shall be protected by ground-fault circuit interrupters. (See 1910.302(b)(3).)

[46 FR 4056, Jan. 16, 1981; 46 FR 40185, Aug. 7, 1981]

Sec. 1910.307 Hazardous (classified) locations.

(a) *Scope.* This section covers the requirements for electric equipment and wiring in locations which are classified depending on the properties of the flammable vapors, liquids or gases, or combustible dusts or fibers which may be present therein and the likelihood that a flammable or combustible concentration or quantity is present. Hazardous (classified) locations may be found in occupancies such as, but not limited to, the following: aircraft hangars, gasoline dispensing and service stations, bulk storage plants for gasoline or other volatile flammable liquids, paint-finishing process plants, health care facilities, agricultural or other facilities where excessive combustible dusts may be present, marinas, boat yards, and petroleum and chemical processing plants. Each room, section or area shall be considered individually in determining its classification. These hazardous (classified) locations are assigned six designations as follows:

Class I, Division 1
Class I, Division 2
Class II, Division 1
Class II, Division 2
Class III, Division 1
Class III, Division 2

For definitions of these locations see 1910.399(a). All applicable requirements in this subpart shall apply to hazardous (classified) locations, unless modified by provisions of this section.

(b) *Electrical Installations.* Equipment, wiring methods, and installations of equipment in hazardous (classified) locations shall be intrinsically safe, approved for the hazardous (classified) location, or safe or for the hazardous (classified) location. Requirements for each of these options are as follows:

(1) *Intrinsically safe.* Equipment and associated wiring approved as intrinsically safe shall be permitted in any hazardous (classified) location for which it is approved.

(2) *Approved for the hazardous (classified) location.* (i) Equipment shall be approved not only for the class of location but also for the ignitible or combustible properties of the specific gas, vapor, dust, or fiber that will be present.

Note: NFPA 70, the National Electrical Code, lists or defines hazardous gases, vapors, and dusts by "Groups" characterized by their ignitible or combustible properties.

(ii) Equipment shall be marked to show the class, group, and operating temperature or temperature range, based on operation in a 40 degrees C ambient, for which it is approved. The temperature marking may not exceed the ignition temperature of the specific gas or vapor to be encountered. However, the following provisions modify this marking requirement for specific equipment:

(A) Equipment of the non-heat-producing type, such as junction boxes, conduit, and fittings, and equipment of the heat-producing type having a maximum temperature not more than 100 degrees C (212 degrees F) need not have a marked operating temperature or temperature range.

(B) Fixed lighting fixtures marked for use in Class I, Division 2 locations only, need not be marked to indicate the group.

(C) Fixed general-purpose equipment in Class I locations, other than lighting fixtures, which is acceptable for use in Class I, Division 2 locations need not be marked with the class, group, division, or operating temperature.

(D) Fixed dust-tight equipment, other than lighting fixtures, which is acceptable for use in Class II, Division 2 and Class III locations need not be marked with the class, group, division, or operating temperature.

(3) Safe for the hazardous (classified) location. Equipment which is safe for the location shall be of a type and design which the employer demonstrates will provide protection from the hazards arising from the combustibility and flammability of vapors, liquids, gases, dusts, or fibers.

Note: The National Electrical Code, NFPA 70, contains guidelines for determining the type and design of equipment and installations which will meet this requirement. The guidelines of this document address electric wiring, equipment, and systems installed in hazardous (classified) locations and contain specific provisions for the following: wiring methods, wiring connections; conductor insulation, flexible cords, sealing and drainage, transformers, capacitors, switches, circuit breakers, fuses, motor controllers, receptacles, attachment plugs, meters, relays, instruments, resistors, generators, motors, lighting fixtures, storage battery charging equipment, electric cranes, electric hoists and similar equipment, utilization equipment, signaling systems, alarm systems, remote control systems, local loud speaker and communication systems, ventilation piping, live parts, lightning surge protection, and grounding. Compliance with these guidelines will constitute one means, but not the only means, of compliance with this paragraph.

(c) **Conduits.** All conduits shall be threaded and shall be made wrench-tight. Where it is impractical to make a threaded joint tight, a bonding jumper shall be utilized.

(d) **Equipment in Division 2 locations.** Equipment that has been approved for a Division 1 location may be installed in a Division 2 location of the same class and group. General-purpose equipment or equipment in general-purpose enclosures may be installed in Division 2 locations if the equipment does not constitute a source of ignition under normal operating conditions.

[46 FR 4056, Jan. 16, 1981; 46 FR 40185, Aug. 7, 1981]

Sec. 1910.308 Special systems.

(a) **Systems over 600 volts, nominal.** Paragraphs (a) (1) through (4) of this section cover the general requirements for all circuits and equipment operated at over 600 volts.

(1) *Wiring methods for fixed installations.* (i) Above-ground conductors shall be installed in rigid metal conduit, in intermediate metal conduit, in cable trays, in cablebus, in other suitable raceways, or as open runs of metal-clad cable suitable for the use and purpose. However, open runs of non-metallic-sheathed cable or of bare conductors or busbars may be installed in locations accessible only to qualified persons. Metallic shielding components, such as tapes, wires, or braids for conductors, shall be grounded. Open runs of insulated wires and cables having a bare lead sheath or a braided outer covering shall be supported in a manner designed to prevent physical damage to the braid or sheath.

(ii) Conductors emerging from the ground shall be enclosed in approved raceways. (See 1910.302(b)(3).)

(2) *Interrupting and isolating devices.* (i) Circuit breaker installations located indoors shall consist of metal-enclosed units or fire-resistant cell-mounted units. In locations accessible only to qualified personnel, open mounting of circuit breakers is permitted. A means of indicating the open and closed position of circuit breakers shall be provided.

(ii) Fused cutouts installed in buildings or tranformer vaults shall be of a type approved for the purpose. They shall be readily accessible for fuse replacement.

(iii) A means shall be provided to completely isolate equipment for inspection and repairs. Isolating means which are not designed to interrupt the load curent of the circuit shall be either interlocked with an approved circuit interrupter or provided with a sign warning against opening them under load.

(3) *Mobile and portable equipment.* (i) *Power cable connections to mobile machines.* A metallic enclosure shall be provided on the mobile machine for enclosing the terminals of the power cable. The enclosure shall include provisions for a solid connection for the ground wire(s) terminal to effectively ground the machine frame. The method of cable termination used shall prevent any strain or pull on the cable from stressing the electrical connections. The enclosure shall have provision for locking so only authorized qualified persons may open it and shall be marked with a sign warning of the presence of energized parts.

(ii) *Guarding live parts.* All energized switching and control parts shall be enclosed in effectively grounded metal cabinets or enclosures. Circuit breakers and protective equipment shall have the operating means projecting through the metal cabinet or enclosure so these units can be reset without locked doors being opened. Enclosures and metal cabinets shall be locked so that only authorized qualified persons have access and shall be marked with a sign warning of the presence of energized parts. Collector ring assemblies on revolving-type machines (shovels, draglines, etc.) shall be guarded.

(4) *Tunnel installation* — (i) *Application.* The provisions of this paragraph apply to installation and use of high-voltage power distribution and utilization equipment which is portable and/or mobile, such as substations, trailers, cars, mobile shovels, draglines, hoists, drills, dredges, compressors, pumps, conveyors, and underground excavators.

(ii) *Conductors.* Conductors in tunnels shall be installed in one or more of the following:

(A) Metal conduit or other metal raceway,

(B) Type MC cable, or

(C) Other approved multiconductor cable.

Conductors shall also be so located or guarded as to protect them from physical damage. Multiconductor portable cable may supply mobile equipment. An equipment grounding conductor shall be run with circuit conductors inside the metal raceway or inside the multiconductor cable jacket. The equipment grounding conductor may be insulated or bare.

(iii) *Guarding live parts.* Bare terminals of transformers, switches, motor controllers, and other equipment shall be enclosed to prevent accidental contact with energized parts. Enclosures for use in tunnels shall be drip-proof, weatherproof, or submersible as required by the environmental conditions.

(iv) *Disconnecting means.* A disconnecting means that simultaneously opens all ungrounded conductors shall be installed at each transformer or motor location.

(v) *Grounding and bonding.* All nonenergized metal parts of electric equipment and metal raceways and cable sheaths shall be effectively grounded and bonded to all metal pipes and rails at the portal and at intervals not exceeding 1000 feet throughout the tunnel.

(b) *Emergency power systems* — (1) *Scope.* The provisions for emergency systems apply to circuits, systems, and equipment intended to supply power for illumination and special loads, in the event of failure of the normal supply.

(2) *Wiring methods.* Emergency circuit wiring shall be kept entirely independent of all other wiring and equipment and may not enter the same raceway, cable, box, or cabinet or other wiring except either where common circuit elements suitable for the purpose are required, or for transferring power from the normal to the emergency source.

(3) *Emergency illumination.* Where emergency lighting is necessary, the system shall be so arranged that the failure of any individual lighting element, such as the burning out of a light bulb, cannot leave any space in total darkness.

(c) *Class 1, Class 2, and Class 3 remote control, signaling, and power-limited circuits* — (1) *Classification.* Class 1, Class 2, or Class 3 remote control, signaling, or power-limited circuits are characterized by their usage and electrical power limitation which differentiates them from light and power circuits. These circuits are classified in accordance with their respective voltage and power limitations as summarized in paragraphs (c)(1)(i) through (c)(1)(iii) of this section.

(i) *Class 1 circuits.* (A) A Class 1 power-limited circuit is supplied from a source having a rated output of not more than 30 volts and 1000 volt-amperes.

(B) A Class 1 remote control circuit or a Class 1 signaling circuit has a voltage which does not exceed 600 volts; however, the power output of the source need not be limited.

(ii) *Class 2 and Class 3 circuits.* (A) Power for Class 2 and Class 3 circuits is limited either inherently (in which no overcurrent protection is required) or by a combination of a power source and overcurrent protection.

(B) The maximum circuit voltage is 150 volts AC or DC for a Class 2 inherently limited power source, and 100 volts AC or DC for a Class 3 inherently limited power source.

(C) The maximum circuit voltage is 30 volts AC and 60 volts DC for a Class 2 power source limited by overcurrent protection, and 150 volts AC or DC for a Class 3 power source limited by overcurrent protection.

(iii) The maximum circuit voltages in paragraphs (c)(1)(i) and (c)(1)(ii) of this section apply to sinusoidal AC or continuous DC power sources, and where wet contact occurence is not likely.

(2) *Marking.* A Class 2 or Class 3 power supply unit shall be durably marked where plainly visible to indicate the class of supply and its electrical rating. (See 1910.302(b)(3).)

(d) *Fire protective signaling systems.* (See Sec. 1910.302(b)(3).)

(1) *Classifications.* Fire protective signaling circuits shall be classified either as non-power limited or power limited.

(2) *Power sources.* The power sources for use with fire protective signaling circuits shall be either power limited or nonlimited as follows:

(i) The power supply of non-power-limited fire protective signaling circuits shall have an output voltage not in excess of 600 volts.

(ii) The power for power-limited fire protective signaling circuits shall be either inherently limited, in which no overcurrent protection is required, or limited by a combination of a power source and overcurrent protection.

(3) *Non-power-limited conductor location.* Non-power-limited fire protective signaling circuits and Class 1 circuits may occupy the same enclosure, cable, or raceway provided all conductors are insulated for maximum voltage of any conductor within the enclosure, cable, or raceway. Power supply and fire protective signaling circuit conductors are permitted in the same enclosure, cable, or raceway only if connected to the same equipment.

(4) *Power-limited conductor location.* Where open conductors are installed, power-limited fire protective signaling circuits shall be separated at least 2 inches from conductors of any light, power, Class 1, and non-power-limited fire protective signaling circuits unless a special and equally protective method of conductor separation is employed. Cables and conductors of two or more power-limited fire protective signaling circuits or Class 3 circuits are permitted in the same cable, enclosure, or raceway. Conductors of one or more Class 2 circuits are permitted within the same cable, enclosure, or raceway with conductors of power-limited fire protective signaling circuits provided that the insulation of Class 2 circuit conductors in the cable, enclosure, or raceway is at least that needed for the power-limited fire protective signaling circuits.

(5) *Identification.* Fire protective signaling circuits shall be identi-

fied at terminal and junction locations in a manner which will prevent unintentional interference with the signaling circuit during testing and servicing. Power-limited fire protective signaling circuits shall be durably marked as such where plainly visible at terminations.

(e) *Communications systems* — (1) *Scope.* These provisions for communication systems apply to such systems as central-station-connected and non-central-station-connected telephone circuits, radio and television receiving and transmitting equipment, including community antenna television and radio distribution systems, telegraph, district messenger, and outside wiring for fire and burglar alarm, and similar central station systems. These installations need not comply with the provisions of 1910.303 through 1910.308(d), except 1910.304(c)(1) and 1910.307(b).

(2) *Protective devices.* (i) Communication circuits so located as to be exposed to accidental contact with light or power conductors operating at over 300 volts shall have each circuit so exposed provided with a protector approved for the purpose.

(ii) Each conductor of a lead-in from an outdoor antenna shall be provided with an antenna discharge unit or other suitable means that will drain static charges from the antenna system.

(3) *Conductor location* — (i) *Outside of buildings.* (a) Receiving distribution lead-in or aerial-drop cables attached to buildings and lead-in conductors to radio transmitters shall be so installed as to avoid the possibility of accidental contact with electric light or power conductors.

(b) The clearance between lead-in conductors and any lightning protection conductors may not be less than 6 feet.

(ii) *On poles.* Where practicable, communication conductors on poles shall be located below the light or power conductors. Communications conductors may not be attached to a crossarm that carries light or power conductors.

(iii) *Inside of buildings.* Indoor antennas, lead-ins, and other communication conductors attached as open conductors to the inside of buildings shall be located at least 2 inches from conductors of any light or power or Class 1 circuits unless a special and equally protective method of conductor separation, approved for the purpose, is employed.

(4) *Equipment location.* Outdoor metal structures supporting antennas, as well as self-supporting antennas such as vertical rods or dipole structures, shall be located as far away from overhead conductors of electric light and power circuits of over 150 volts to ground as necessary to avoid the possibility of the antenna or structure falling into or making accidental contact with such circuits.

(5) *Grounding* — (i) *Lead-in conductors.* If exposed to contact with electric light and power conductors, the metal sheath of aerial cables entering buildings shall be grounded or shall be interrupted close to the entrance to the building by an insulating joint or equivalent device. Where protective devices are used, they shall be grounded in an approved manner.

(ii) *Antenna structures.* Masts and metal structures supporting antennas shall be permanently and effectively grounded without splice or connection in the grounding conductor.

(iii) *Equipment enclosures.* Transmitters shall be enclosed in a metal frame or grill or separated from the operating space by a barrier, all metallic parts of which are effectively connected to ground. All external metal handles and controls accessible to the operating personnel shall be effectively grounded. Unpowered equipment and enclosures shall be considered grounded where connected to an attached coaxial cable with an effectively grounded metallic shield.

[46 FR 4056, Jan. 16, 1981; 46 FR 40185, Aug. 7, 1981]

Definitions

Sec. 1910.399 Definitions.

(a) *Definitions applicable to Secs. 1910.302 through 1910.330* — (1) *Acceptable.* An installation or equipment is acceptable to the Assistant Secretary of Labor, and approved within the meaning of this Subpart S:

(i) If it is accepted, or certified, or listed, or labeled, or otherwise determined to be safe by a nationally recognized testing laboratory; or

(ii) With respect to an installation or equipment of a kind which no nationally recognized testing laboratory accepts, certifies, lists, labels, or determines to be safe, if it is inspected or tested by another Federal agency, or by a State, municipal, or other local authority responsible for enforcing occupational safety provisions of the National Electrical Code and found in compliance with the provisions of the National Electrical Code as applied in this Subpart; or

(iii) With respect to custom-made equipment or related installations which are designed, fabricated for, and intended for use by a particular customer, if it is determined to be safe for its intended use by its manufacturer on the basis of test data which the employer keeps and makes available for inspection to the Assistant Secretary and his authorized representatives. Refer to 1910.7 for definition of nationally recognized testing laboratory.

(2) *Accepted.* An installation is "accepted" if it has been inspected and found by a nationally recognized testing laboratory to conform to specified plans or to procedures of applicable codes.

(3) *Accessible.* (As applied to wiring methods.) Capable of being removed or exposed without damaging the building structure or finish, or not permanently closed in by the structure or finish of the building. (See "concealed" and "exposed.")

(4) *Accessible.* (As applied to equipment.) Admitting close approach; not guarded by locked doors, elevation, or other effective means. (See "Readily accessible.")

(5) *Ampacity.* Current-carrying capacity of electric conductors expressed in amperes.

(6) *Appliances.* Utilization equipment, generally other than industrial, normally built in standardized sizes or types, which is installed or connected as a unit to perform one or more functions such as clothes washing, air conditioning, food mixing, deep frying, etc.

(7) *Approved.* Acceptable to the authority enforcing this subpart. The authority enforcing this subpart is the Assistant Secretary of Labor for Occupational Safety and Health. The definition of "acceptable" indicates what is acceptable to the Assistant Secretary of Labor, and therefore approved within the meaning of this Subpart.

(8) *Approved for the purpose.* Approved for a specific purpose, environment, or application described in a particular standard requirement.

Suitability of equipment or materials for a specific purpose, environment or application may be determined by a nationally recognized testing laboratory, inspection agency or other organization concerned with product evaluation as part of its listing and labeling program. (See "Labeled" or "Listed.")

(9) *Armored cable.* Type AC armored cable is a fabricated assembly of insulated conductors in a flexible metallic enclosure.

(10) *Askarel.* A generic term for a group of nonflammable synthetic chlorinated hydrocarbons used as electrical insulating media. Askarels of various compositional types are used. Under arcing conditions the gases produced, while consisting predominantly of noncombustible hydrogen chloride, can include varying amounts of combustible gases depending upon the askarel type.

(11) *Attachment plug (Plug cap) (Cap).* A device which, by insertion in a receptacle, establishes connection between the conductors of the attached flexible cord and the conductors connected permanently to the receptacle.

(12) *Automatic.* Self-acting, operating by its own mechanism when actuated by some impersonal influence, as, for example, a change in current strength, pressure, temperature, or mechanical configuration.

(13) *Bare conductor.* See "Conductor."

(14) *Bonding.* The permanent joining of metallic parts to form an electrically conductive path which will assure electrical continuity and the capacity to conduct safely any current likely to be imposed.

(15) *Bonding jumper.* A reliable conductor to assure the required electrical conductivity between metal parts required to be electrically connected.

(16) *Branch circuit.* The circuit conductors between the final overcurrent device protecting the circuit and the outlet(s).

(17) *Building.* A structure which stands alone or which is cut off from adjoining structures by fire walls with all openings therein protected by approved fire doors.

(18) *Cabinet.* An enclosure designed either for surface or flush mounting, and provided with a frame, mat, or trim in which a swinging door or doors are or may be hung.

(19) *Cable tray system.* A cable tray system is a unit or assembly of units or sections, and associated fittings, made of metal or other noncombustible materials forming a rigid structural system used to support cables. Cable tray systems include ladders, troughs, channels, solid bottom trays, and other similar structures.

(20) *Cablebus.* Cablebus is an approved assembly of insulated conductors with fittings and conductor terminations in a completely enclosed, ventilated, protective metal housing.

(21) *Center pivot irrigation machine.* A center pivot irrigation machine is a multi-motored irrigation machine which revolves around a central pivot and employs alignment switches or similar devices to control individual motors.

(22) *Certified.* Equipment is "certified" if it (a) has been tested and found by a nationally recognized testing laboratory to meet nationally recognized standards or to be safe for use in a specified manner, or (b) is of a kind whose production is periodically inspected by a nationally recognized testing laboratory, and (c) it bears a label, tag, or other record of certification.

(23) *Circuit breaker.* (i) *(600 volts nominal, or less).* A device designed to open and close a circuit by nonautomatic means and to open the circuit automatically on a predetermined overcurrent without injury to itself when properly applied within its rating.

(ii) *(Over 600 volts, nominal).* A switching device capable of making, carrying, and breaking currents under normal circuit conditions, and also making, carrying for a specified time, and breaking currents under specified abnormal circuit conditions, such as those of short circuit.

(24) *Class I locations.* Class I locations are those in which flammable gases or vapors are or may be present in the air in quantities sufficient to produce explosive or ignitible mixtures. Class I locations include the following:

(i) *Class I, Division 1.* A Class I, Division 1 location is a location: (a) in which hazardous concentrations of flammable gases or vapors may exist under normal operating conditions; or (b) in which hazardous concentrations of such gases or vapors may exist frequently because of repair or maintenance operations or because of leakage; or (c) in which breakdown or faulty operation of equipment or processes might release hazardous concentrations of flammable gases or vapors, and might also cause simultaneous failure of electric equipment.

Note: This classification usually includes locations where volatile flammable liquids or liquefied flammable gases are transferred from one container to another; interiors of spray booths and areas in the vicinity of spraying and painting operations where volatile flammable solvents are used; locations containing open tanks or vats of volatile flammable liquids; drying rooms or compartments for the evaporation of flammable solvents; locations containing fat and oil extraction equipment using volatile flammable solvents; portions of cleaning and dyeing plants where flammable liquids are used; gas generator rooms and other portions of gas manufacturing plants where flammable gas may escape; inadequately ventilated pump rooms for flammable gas or for volatile flammable liquids; the interiors of refrigerators and freezers in which volatile flammable materials are stored in open, lightly stoppered, or easily ruptured containers; and all other locations where ignitible concentrations of flammable vapors or gases are likely to occur in the course of normal operations.

(ii) *Class I, Division 2.* A Class I, Division 2 location is a location: (a) in which volatile flammable liquids or flammable gases are handled, processed, or used, but in which the hazardous liquids, vapors, or gases will normally be confined within closed containers or closed systems from which they can escape only in case of accidental rupture or breakdown of such containers or systems, or in case of abnormal operation of equipment; or (b) in which hazardous concentrations of gases or vapors are normally prevented by positive mechanical ventilation, and which might become hazardous through failure or abnormal operations of the ventilating equipment; or (c) that is adjacent to a Class I, Division 1 location, and to which hazardous concentrations of gases or vapors might occasionally be communicated unless such communication is prevented by adequate positive-pressure ventilation from a source of clean air, and effective safeguards against ventilation failure are provided.

Note: This classification usually includes locations where volatile flammable liquids or flammable gases or vapors are used, but which would become hazardous only in case of an accident or of some unusual operating condition. The quantity of flammable material that might escape in case of accident, the adequacy of ventilating equipment, the total area involved, and the record of the industry or business with respect to explosions or fires are all factors that merit consideration in determining the classification and extent of each location.

Piping without valves, checks, meters, and similar devices would not ordinarily introduce a hazardous condition even though used for flammable liquids or gases. Locations used for the storage of flammable liquids or a liquefied or compressed gases in sealed containers would not normally be considered hazardous unless also subject to other hazardous conditions.

Electrical conduits and their associated enclosures separated from process fluids by a single seal or barrier are classed as a Division 2 location if the outside of the conduit and enclosures is a nonhazardous location.

(25) *Class II locations.* Class II locations are those that are hazardous because of the presence of combustible dust. Class II locations include the following:

(i) *Class II, Division 1.* A Class II, Division 1 location is a location: (a) In which combustible dust is or may be in suspension in the air under normal operating conditions, in quantities sufficient to produce explosive or ignitible mixtures; or (b) where mechanical failure or abnormal operation of machinery or equipment might cause such explosive or ignitible mixtures to be produced, and might also provide a source of ignition through simultaneous failure of electric equipment, operation of protection devices, or from other causes, or (c) in which combustible dusts of an electrically conductive nature may be present.

Note: This classification may include areas of grain handling and processing plants, starch plants, sugar-pulverizing plants, malting plants, hay-grinding plants, coal pulverizing plants, areas where metal dusts and powders are produced or processed, and other similar locations which contain dust producing machinery and equipment (except where the equipment is dust-tight or vented to the outside). These areas would have combustible dust in the air, under normal operating conditions, in quantities sufficient to produce explosive or ignitible mixtures. Combustible dusts which are electrically nonconductive include dusts produced in the handling and processing of grain and grain products, pulverized sugar and cocoa, dried egg and milk powders, pulverized spices, starch and pastes, potato and woodflour, oil meal from beans and seed, dried hay, and other organic materials which may produce combustible dusts when processed or handled. Dusts containing magnesium or aluminum are particularly hazardous and the use of extreme caution is necessary to avoid ignition and explosion.

(ii) *Class II, Division 2.* A Class II, Division 2 location is a location in which: (a) combustible dust will not normally be in suspension in the air in quantities sufficient to produce explosive or ignitible mixtures, and dust accumulations are normally insufficient to interfere with the normal operation of electrical equipment or other apparatus; or (b) dust may be in suspension in the air as a result of infrequent malfunctioning of handling or processing equipment, and dust accumulations resulting therefrom may be ignitible by abnormal operation or failure of electrical equipment or other apparatus.

Note: This classification includes locations where dangerous concentrations of suspended dust would not be likely but where dust accumulations might form on or in the vicinity of electric equipment. These areas may contain equipment from which appreciable quantities of dust would escape under abnormal operating conditions or be adjacent to a Class II Division 1 location, as described above, into which an explosive or ignitible concentration of dust may be put into suspension under abnormal operating conditions.

(26) *Class III locations.* Class III locations are those that are hazardous because of the presence of easily ignitible fibers or flyings but in which such fibers or flyings are not likely to be in suspension in the air in quantities sufficient to produce ignitible mixtures. Class III locations include the following:

(i) *Class III, Division 1.* A Class III, Division 1 location is a location in which easily ignitible fibers or materials producing combustible flyings are handled, manufactured, or used.

Note: Such locations usually include some parts of rayon, cotton, and other textile mills; combustible fiber manufacturing and processing plants; cotton gins and cotton-seed mills; flax-processing plants; clothing manufacturing plants; woodworking plants, and establishments; and industries involving similar hazardous processes or conditions.

Easily ignitible fibers and flyings include rayon, cotton (including cotton linters and cotton waste), sisal or henequen, istle, jute, hemp, tow, cocoa fiber, oakum, baled waste kapok, Spanish moss, excelsior, and other materials of similar nature.

(ii) *Class III, Division 2.* A Class III, Division 2 location is a location in which easily ignitible fibers are stored or handled, except in process of manufacture.

(27) *Collector ring.* A collector ring is an assembly of slip rings for transferring electrical energy from a stationary to a rotating member.

(28) *Concealed.* Rendered inaccessible by the structure or finish of the building. Wires in concealed raceways are considered concealed, even though they may become accessible by withdrawing them. [See "Accessible. (As applied to wiring methods.)"]

(29) *Conductor.* (i) Bare. A conductor having no covering or electrical insulation whatsoever.

(ii) *Covered.* A conductor encased within material of composition or thickness that is not recognized as electrical insulation.

(iii) *Insulated.* A conductor encased within material of composition and thickness that is recognized as electrical insulation.

(30) *Conduit body.* A separate portion of a conduit or tubing system that provides access through a removable cover(s) to the interior of the system at a junction of two or more sections of the system or at a terminal point of the system. Boxes such as FS and FD or larger cast or sheet metal boxes are not classified as conduit bodies.

(31) *Controller.* A device or group of devices that serves to govern, in some predetermined manner, the electric power delivered to the apparatus to which it is connected.

(32) *Cooking unit, counter-mounted.* A cooking appliance designed for mounting in or on a counter and consisting of one or more heating elements, internal wiring, and built-in or separately mountable controls. (See "Oven, wall-mounted.")

(33) *Covered conductor.* See "Conductor."

(34) *Cutout.* (Over 600 volts, nominal.) An assembly of a fuse support with either a fuseholder, fuse carrier, or disconnecting blade. The fuseholder or fuse carrier may include a conducting element (fuse link), or may act as the disconnecting blade by the inclusion of a nonfusible member.

(35) *Cutout box.* An enclosure designed for surface mounting and

having swinging doors or covers secured directly to and telescoping with the walls of the box proper. (See "Cabinet.")

(36) *Damp location.* See "Location."

(37) *Dead front.* Without live parts exposed to a person on the operating side of the equipment.

(38) *Device.* A unit of an electrical system which is intended to carry but not utilize electric energy.

(39) *Dielectric heating.* Dielectric heating is the heating of a nominally insulating material due to its own dielectric losses when the material is placed in a varying electric field.

(40) *Disconnecting means.* A device, or group of devices, or other means by which the conductors of a circuit can be disconnected from their source of supply.

(41) *Disconnecting (or Isolating) switch.* (Over 600 volts, nominal.) A mechanical switching device used for isolating a circuit or equipment from a source of power.

(42) *Dry location.* See "Location."

(43) *Electric sign.* A fixed, stationary, or portable self-contained, electrically illuminated utilization equipment with words or symbols designed to convey information or attract attention.

(44) *Enclosed.* Surrounded by a case, housing, fence or walls which will prevent persons from accidentally contacting energized parts.

(45) *Enclosure.* The case or housing of apparatus, or the fence or walls surrounding an installation to prevent personnel from accidentally contacting energized parts, or to protect the equipment from physical damage.

(46) *Equipment.* A general term including material, fittings, devices, appliances, fixtures, apparatus, and the like, used as a part of, or in connection with, an electrical installation.

(47) *Equipment grounding conductor.* See "Grounding conductor, equipment."

(48) *Explosion-proof apparatus.* Apparatus enclosed in a case that is capable of withstanding an explosion of a specified gas or vapor which may occur within it and of preventing the ignition of a specified gas or vapor surrounding the enclosure by sparks, flashes, or explosion of the gas or vapor within, and which operates at such an external temperature that it will not ignite a surrounding flammable atmosphere.

(49) *Exposed.* (As applied to live parts.) Capable of being inadvertently touched or approached nearer than a safe distance by a person. It is applied to parts not suitably guarded, isolated, or insulated. (See "Accessible." and "Concealed.")

(50) *Exposed.* (As applied to wiring methods.) On or attached to the surface or behind panels designed to allow access. [See "Accessible. (As applied to wiring methods.)"]

(51) *Exposed.* (For the purposes of 1910.308(e), Communications systems.) Where the circuit is in such a position that in case of failure of supports or insulation, contact with another circuit may result.

(52) *Externally operable.* Capable of being operated without exposing the operator to contact with live parts.

(53) *Feeder.* All circuit conductors between the service equipment, or the generator switchboard of an isolated plant, and the final branch-circuit overcurrent device.

(54) *Fitting.* An accessory such as a locknut, bushing, or other part of a wiring system that is intended primarily to perform a mechanical rather than an electrical function.

(55) *Fuse.* (Over 600 volts, nominal.) An overcurrent protective device with a circuit opening fusible part that is heated and severed by the passage of overcurrent through it. A fuse comprises all the parts that form a unit capable of performing the prescribed functions. It may or may not be the complete device necessary to connect it into an electrical circuit.

(56) *Ground.* A conducting connection, whether intentional or accidental, between an electrical circuit or equipment and the earth, or to some conducting body that serves in place of the earth.

(57) *Grounded.* Connected to earth or to some conducting body that serves in place of the earth.

(58) *Grounded, effectively.* (Over 600 volts, nominal.) Permanently connected to earth through a ground connection of sufficiently low impedance and having sufficient ampacity that ground fault current which may occur cannot build up to voltages dangerous to personnel.

(59) *Grounded conductor.* A system or circuit conductor that is intentionally grounded.

(60) *Grounding conductor.* A conductor used to connect equipment or the grounded circuit of a wiring system to a grounding electrode or electrodes.

(61) *Grounding conductor, equipment.* The conductor used to connect the non-current-carrying metal parts of equipment, raceways, and other enclosures to the system grounded conductor and/or the grounding electrode conductor at the service equipment or at the source of a separately derived system.

(62) *Grounding electrode conductor.* The conductor used to connect the grounding electrode to the equipment grounding conductor and/or

to the grounded conductor of the circuit at the service equipment or at the source of a separately derived system.

(63) *Ground-fault circuit-interrupter.* A device whose function is to interrupt the electric circuit to the load when a fault current to ground exceeds some predetermined value that is less than that required to operate the overcurrent protective device of the supply circuit.

(64) *Guarded.* Covered, shielded, fenced, enclosed, or otherwise protected by means of suitable covers, casings, barriers, rails, screens, mats, or platforms to remove the likelihood of approach to a point of danger or contact by persons or objects.

(65) *Health care facilities.* Buildings or portions of buildings and mobile homes that contain, but are not limited to, hospitals, nursing homes, extended care facilities, clinics, and medical and dental offices, whether fixed or mobile.

(66) *Heating equipment.* For the purposes of 1910.306(g), the term "heating equipment" includes any equipment used for heating purposes if heat is generated by induction or dielectric methods.

(67) *Hoistway.* Any shaftway, hatchway, well hole, or other vertical opening or space in which an elevator or dumbwaiter is designed to operate.

(68) *Identified.* Identified, as used in reference to a conductor or its terminal, means that such conductor or terminal can be readily recognized as grounded.

(69) *Induction heating.* Induction heating is the heating of a nominally conductive material due to its own $I\backslash2\backslash R$ losses when the material is placed in a varying electromagnetic field.

(70) *Insulated conductor.* See "Conductor."

(71) *Interrupter switch.* (Over 600 volts, nominal.) A switch capable of making, carrying, and interrupting specified currents.

(72) *Irrigation machine.* An irrigation machine is an electrically driven or controlled machine, with one or more motors, not hand portable, and used primarily to transport and distribute water for agricultural purposes.

(73) *Isolated.* Not readily accessible to persons unless special means for access are used.

(74) *Isolated power system.* A system comprising an isolating transformer or its equivalent, a line isolation monitor, and its ungrounded circuit conductors.

(75) *Labeled.* Equipment is "labeled" if there is attached to it a label, symbol, or other identifying mark of a nationally recognized testing laboratory which, (a) makes periodic inspections of the production of

such equipment, and (b) whose labeling indicates compliance with nationally recognized standards or tests to determine safe use in a specified manner.

(76) *Lighting outlet.* An outlet intended for the direct connection of a lampholder, a lighting fixture, or a pendant cord terminating in a lampholder.

(77) *Listed.* Equipment is "listed" if it is of a kind mentioned in a list which, (a) is published by a nationally recognized laboratory which makes periodic inspection of the production of such equipment, and (b) states such equipment meets nationally recognized standards or has been tested and found safe for use in a specified manner.

(78) *Location* — (i) *Damp location.* Partially protected locations under canopies, marquees, roofed open porches, and like locations, and interior locations subject to moderate degrees of moisture, such as some basements, some barns, and some cold-storage warehouses.

(ii) *Dry location.* A location not normally subject to dampness or wetness. A location classified as dry may be temporarily subject to dampness or wetness, as in the case of a building under construction.

(iii) *Wet location.* Installations underground or in concrete slabs or masonry in direct contact with the earth, and locations subject to saturation with water or other liquids, such as vehicle-washing areas, and locations exposed to weather and unprotected.

(79) *Medium voltage cable.* Type MV medium voltage cable is a single or multiconductor solid dielectric insulated cable rated 2000 volts or higher.

(80) *Metal-clad cable.* Type MC cable is a factory assembly of one or more conductors, each individually insulated and enclosed in a metallic sheath of interlocking tape, or a smooth or corrugated tube.

(81) *Mineral-insulated metal-sheathed cable.* Type MI mineral-insulated metal-sheathed cable is a factory assembly of one or more conductors insulated with a highly compressed refractory mineral insulation and enclosed in a liquidtight and gastight continuous copper sheath.

(82) *Mobile X-ray.* X-ray equipment mounted on a permanent base with wheels and/or casters for moving while completely assembled.

(83) *Nonmetallic-sheathed cable.* Nonmetallic-sheathed cable is a factory assembly of two or more insulated conductors having an outer sheath of moisture resistant, flame-retardant, nonmetallic material. Nonmetallic sheathed cable is manufactured in the following types:

(i) *Type NM.* The overall covering has a flame-retardant and moisture-resistant finish.

(ii) *Type NMC*. The overall covering is flame-retardant, moisture-resistant, fungus-resistant, and corrosion-resistant.

(84) *Oil (filled) cutout*. (Over 600 volts, nominal.) A cutout in which all or part of the fuse support and its fuse link or disconnecting blade are mounted in oil with complete immersion of the contacts and the fusible portion of the conducting element (fuse link), so that arc interruption by severing of the fuse link or by opening of the contacts will occur under oil.

(85) *Open wiring on insulators*. Open wiring on insulators is an exposed wiring method using cleats, knobs, tubes, and flexible tubing for the protection and support of single insulated conductors run in or on buildings, and not concealed by the building structure.

(86) *Outlet*. A point on the wiring system at which current is taken to supply utilization equipment.

(87) *Outline lighting*. An arrangement of incandescent lamps or electric discharge tubing to outline or call attention to certain features such as the shape of a building or the decoration of a window.

(88) *Oven, wall-mounted*. An oven for cooking purposes designed for mounting in or on a wall or other surface and consisting of one of more heating elements, internal wiring, and built-in or separately mountable controls. (See "Cooking unit, counter-mounted.")

(89) *Overcurrent*. Any current in excess of the rated current of equipment or the ampacity of a conductor. It may result from overload (see definition), short circuit, or ground fault. A current in excess of rating may be accommodated by certain equipment and conductors for a given set of conditions. Hence the rules for overcurrent protection are specific for particular situations.

(90) *Overload*. Operation of equipment in excess of normal, full load rating, or of a conductor in excess of rated ampacity which, when it persists for a sufficient length of time, would cause damage or dangerous overheating. A fault, such as a short circuit or ground fault, is not an overload. (See "Overcurrent.")

(91) *Panelboard*. A single panel or group of panel units designed for assembly in the form of a single panel; including buses, automatic overcurrent devices, and with or without switches for the control of light, heat, or power circuits; designed to be placed in a cabinet or cutout box placed in or against a wall or partition and accessible only from the front. (See "Switchboard.")

(92) *Permanently installed decorative fountains and reflection pools*. Those that are constructed in the ground, on the ground, or in a building in such a manner that the pool cannot be readily disassembled for

storage and are served by electrical circuits of any nature. These units are primarily constructed for their aesthetic value and not intended for swimming or wading.

(93) *Permanently installed swimming pools, wading and therapeutic pools.* Those that are constructed in the ground, on the ground, or in a building in such a manner that the pool cannot be readily disassembled for storage whether or not served by electrical circuits of any nature.

(94) *Portable X-ray.* X-ray equipment designed to be hand-carried.

(95) *Power and control tray cable.* Type TC power and control tray cable is a factory assembly of two or more insulated conductors, with or without associated bare or covered grounding conductors under a nonmetallic sheath, approved for installation in cable trays, in raceways, or where supported by a messenger wire.

(96) *Power fuse.* (Over 600 volts, nominal.) See "Fuse."

(97) *Power-limited tray cable.* Type PLTC nonmetallic-sheathed power limited tray cable is a factory assembly of two or more insulated conductors under a nonmetallic jacket.

(98) *Power outlet.* An enclosed assembly which may include receptacles, circuit breakers, fuseholders, fused switches, buses and watt-hour meter mounting means; intended to supply and control power to mobile homes, recreational vehicles or boats, or to serve as a means for distributing power required to operate mobile or temporarily installed equipment.

(99) *Premises wiring system.* That interior and exterior wiring, including power, lighting, control, and signal circuit wiring together with all of its associated hardware, fittings, and wiring devices, both permanently and temporarily installed, which extends from the load end of the service drop, or load end of the service lateral conductors to the outlet(s). Such wiring does not include wiring internal to appliances, fixtures, motors, controllers, motor control centers, and similar equipment.

(100) *Qualified person.* One familiar with the construction and operation of the equipment and the hazards involved.

(101) *Raceway.* A channel designed expressly for holding wires, cables, or busbars, with additional functions as permitted in this subpart. Raceways may be of metal or insulating material, and the term includes rigid metal conduit, rigid nonmetallic conduit, intermediate metal conduit, liquidtight flexible metal conduit, flexible metallic tubing, flexible metal conduit, electrical metallic tubing, underfloor raceways, cellular concrete floor raceways, cellular metal floor raceways, surface raceways, wireways, and busways.

(102) *Readily accessible.* Capable of being reached quickly for operation, renewal, or inspections, without requiring those to whom ready access is requisite to climb over or remove obstacles or to resort to portable ladders, chairs, etc. (See "Accessible.")

(103) *Receptacle.* A receptacle is a contact device installed at the outlet for the connection of a single attachment plug. A single receptacle is a single contact device with no other contact device on the same yoke. A multiple receptacle is a single device containing two or more receptacles.

(104) *Receptacle outlet.* An outlet where one or more receptacles are installed.

(105) *Remote-control circuit.* Any electric circuit that controls any other circuit through a relay or an equivalent device.

(106) *Sealable equipment.* Equipment enclosed in a case or cabinet that is provided with a means of sealing or locking so that live parts cannot be made accessible without opening the enclosure. The equipment may or may not be operable without opening the enclosure.

(107) *Separately derived system.* A premises wiring system whose power is derived from generator, transformer, or converter winding and has no direct electrical connection, including a solidly connected grounded circuit conductor, to supply conductors originating in another system.

(108) *Service.* The conductors and equipment for delivering energy from the electricity supply system to the wiring system of the premises served.

(109) *Service cable.* Service conductors made up in the form of a cable.

(110) *Service conductors.* The supply conductors that extend from the street main or from transformers to the service equipment of the premises supplied.

(111) *Service drop.* The overhead service conductors from the last pole or other aerial support to and including the splices, if any, connecting to the service-entrance conductors at the building or other structure.

(112) *Service-entrance cable.* Service-entrance cable is a single conductor or multiconductor assembly provided with or without an overall covering, primarily used for services and of the following types:

(i) *Type SE,* having a flame-retardant, moisture-resistant covering, but not required to have inherent protection against mechanical abuse.

(ii) *Type USE,* recognized for underground use, having a moisture-resistant covering, but not required to have a flame-retardant covering or inherent protection against mechanical abuse. Single-conductor cables having an insulation specifically approved for the purpose do not require an outer covering.

(113) *Service-entrance conductors, overhead system.* The service conductors between the terminals of the service equipment and a point usually outside the building, clear of building walls, where joined by tap or splice to the service drop.

(114) *Service entrance conductors, underground system.* The service conductors between the terminals of the service equipment and the point of connection to the service lateral. Where service equipment is located outside the building walls, there may be no service-entrance conductors, or they may be entirely outside the building.

(115) *Service equipment.* The necessary equipment, usually consisting of a circuit breaker or switch and fuses, and their accessories, located near the point of entrance of supply conductors to a building or other structure, or an otherwise defined area, and intended to consititute the main control and means of cutoff of the supply.

(116) *Service raceway.* The raceway that encloses the service-entrance conductors.

(117) *Shielded nonmetallic-sheathed cable.* Type SNM, shielded nonmetallic-sheathed cable is a factory assembly of two or more insulated conductors in an extruded core of moisture-resistant, flame-resistant nonmetallic material, covered with an overlapping spiral metal tape and wire shield and jacketed with an extruded moisture-, flame-, oil-, corrosion-, fungus-, and sunlight-resistant nonmetallic material.

(118) *Show window.* Any window used or designed to be used for the display of goods or advertising material, whether it is fully or partly enclosed or entirely open at the rear and whether or not it has a platform raised higher than the street floor level.

(119) *Sign.* See "Electric Sign."

(120) *Signaling circuit.* Any electric circuit that energizes signaling equipment.

(121) *Special permission.* The written consent of the authority having jurisdiction.

(122) *Storable swimming or wading pool.* A pool with a maximum dimension of 15 feet and a maximum wall height of 3 feet and is so constructed that it may be readily disassembled for storage and reassembled to its original integrity.

(123) *Switchboard.* A large single panel, frame, or assembly of panels which have switches, buses, instruments, overcurrent and other protective devices mounted on the face or back or both. Switchboards are generally accessible from the rear as well as from the front and are not

intended to be installed in cabinets. (See "Panelboard.")

(124) *Switches.*

(i) *General-use switch.* A switch intended for use in general distribution and branch circuits. It is rated in amperes, and it is capable of interrupting its rated current at its rated voltage.

(ii) *General-use snap switch.* A form of general-use switch so constructed that it can be installed in flush device boxes or on outlet box covers, or otherwise used in conjunction with wiring systems recognized by this subpart.

(iii) *Isolating switch.* A switch intended for isolating an electric circuit from the source of power. It has no interrupting rating, and it is intended to be operated only after the circuit has been opened by some other means.

(iv) *Motor-circuit switch.* A switch, rated in horsepower, capable of interrupting the maximum operating overload current of a motor of the same horsepower rating as the switch at the rated voltage.

(125) *Switching devices.* (Over 600 volts, nominal.) Devices designed to close and/or open one or more electric circuits. Included in this category are circuit breakers, cutouts, disconnecting (or isolating) switches, disconnecting means, interrupter switches, and oil (filled) cutouts.

(126) *Transportable X-ray.* X-ray equipment installed in a vehicle or that may readily be disassembled for transport in a vehicle.

(127) *Utilization equipment.* Utilization equipment means equipment which utilizes electric energy for mechanical, chemical, heating, lighting, or similar useful purpose.

(128) *Utilization system.* A utilization system is a system which provides electric power and light for employee workplaces, and includes the premises wiring system and utilization equipment.

(129) *Ventilated.* Provided with a means to permit circulation of air sufficient to remove an excess of heat, fumes, or vapors.

(130) *Volatile flammable liquid.* A flammable liquid having a flash point below 38 degrees C (100 degrees F) or whose temperature is above its flash point.

(131) *Voltage (of a circuit).* The greatest root-mean-square (effective) difference of potential between any two conductors of the circuit concerned.

(132) *Voltage, nominal.* A nominal value assigned to a circuit or system for the purpose of conveniently designating its voltage class (as 120/240, 480Y/277, 600, etc.). The actual voltage at which a circuit operates can vary from the nominal within a range that permits

satisfactory operation of equipment.

(133) *Voltage to ground.* For grounded circuits, the voltage between the given conductor and that point or conductor of the circuit that is grounded; for ungrounded circuits, the greatest voltage between the given conductor and any other conductor of the circuit.

(134) *Watertight.* So constructed that moisture will not enter the enclosure.

(135) *Weatherproof.* So constructed or protected that exposure to the weather will not interfere with successful operation. Rainproof, raintight, or watertight equipment can fulfill the requirements for weatherproof where varying weather conditions other than wetness, such as snow, ice, dust, or temperature extremes, are not a factor.

(136) *Wet location.* See "Location."

(137) *Wireways.* Wireways are sheet-metal troughs with hinged or removable covers for housing and protecting electric wires and cable and in which conductors are laid in place after the wireway has been installed as a complete system.

[46 FR 4056, Jan. 16, 1981; 46 FR 40185, Aug. 7, 1981, as amended at 53 FR 12123, Apr. 12, 1988]

Part 1910, Subpt. S, App. A

Appendix A to Subpart S — Reference Documents

The following references provide information which can be helpful in understanding and complying with the requirements contained in Subpart S:

ANSI A17.1 - 71	Safety Code for Elevators, Dumbwaiters, Escalators and Moving Walks.
ANSI B9.1 - 71	Safety Code for Mechanical Refrigeration.
ANSI B30.2 - 76	Safety Code for Overhead and Gantry Cranes.
ANSI B30.3 - 75	Hammerhead Tower Cranes.
ANSI B30.4 - 73	Safety Code for Portal, Tower, and Pillar Cranes.
ANSI B30.5 - 68	Safety Code for Crawler, Locomotive, and Truck Cranes.
ANSI B30.6 - 77	Derricks.
ANSI B30.7 - 77	Base Mounted Drum Hoists.
ANSI B30.8 - 71	Safety Code for Floating Cranes and Floating Derricks.
ANSI B30.11 - 73	Monorail Systems and Underhung Cranes.
ANSI B30.12 - 75	Handling Loads Suspended from Rotorcraft.

ANSI B30.13 - 77	Controlled Mechanical Storage Cranes.
ANSI B30.15 - 73	Safety Code for Mobile Hydraulic Cranes.
ANSI B30.16 - 73	Overhead Hoists.
ANSI C2 - 81	National Electrical Safety Code.
ANSI C33.27 - 74	Safety Standard for Outlet Boxes and Fittings for Use in Hazardous Locations, Class I, Groups A, B, C, and D, and Class II, Groups E, F, and G.
ANSI K61.1 - 72	Safety Requirements for the Storage and Handling of Anhydrous Ammonia.
ASTM D2155 - 66	Test Method for Autoignition Temperature of Liquid Petroleum Products.
ASTM D3176 - 74	Method for Ultimate Analysis of Coal and Coke.
ASTM D3180 - 74	Method for Calculating Coal and Coke Analyses from As Determined to Different Bases.
IEEE 463 - 77	Standard for Electrical Safety Practices in Electrolytic Cell Line Working Zones.
NFPA 20 - 76	Standard for the Installation of Centrifugal Fire Pumps.
NFPA 30 - 78	Flammable and Combustible Liquids Code.
NFPA 32 - 74	Standard for Drycleaning Plants.
NFPA 33 - 73	Standard for Spray Application Using Flammable and Combustible Materials.
NFPA 34 - 74	Standard for Dip Tanks Containing Flammable or Combustible Liquids.
NFPA 35 - 76	Standard for the Manufacture of Organic Coatings.
NFPA 36 - 74	Standard for Solvent Extraction Plants.
NFPA 40 - 74	Standard for the Storage and Handling of Cellulose Nitrate Motion Picture Film.
NFPA 56A - 73	Standard for the Use of Inhalation Anesthetics (Flammable and Nonflammable).
NFPA 56F - 74	Standard for Nonflammable Medical Gas Systems.
NFPA 58 - 76	Standard for the Storage and Handling of Liquefied Petroleum Gases.
NFPA 59 - 76	Standard for the Storage and Handling of Liquefied Petroleum Gases at Utility Gas Plants.
NFPA 70 - 78	National Electrical Code.
NFPA 70C - 74	Hazardous Locations Classification.
NFPA 70E	Standard for the Electrical Safety Requirements for Employee Workplaces.

162

NFPA 71 - 77	Standard for the Installation, Maintenance, and Use of Central Station Signaling Systems.
NFPA 72A - 75	Standard for the Installation, Maintenance, and Use of Local Protective Signaling Systems for Watchman, Fire Alarm, and Supervisory Service.
NFPA 72B - 75	Standard for the Installation, Maintenance, and Use of Auxiliary Protective Signaling Systems for Fire Alarm Service.
NFPA 72C - 75	Standard for the Installation, Maintenance, and Use of Remote Station Protective Signaling Systems.
NFPA 72D - 75	Standard for the Installation, Maintenance, and Use of Proprietary Protective Signaling Systems for Watchman, Fire Alarm, and Supervisory Service.
NFPA 72E - 74	Standard for Automatic Fire Detectors.
NFPA 74 - 75	Standard for Installation, Maintenance, and Use of Household Fire Warning Equipment.
NFPA 76A - 73	Standard for Essential Electrical Systems for Health Care Facilities.
NFPA 77 - 72	Recommended Practice on Static Electricity.
NFPA 80 - 77	Standard for Fire Doors and Windows.
NFPA 86A - 73	Standard for Ovens and Furnaces; Design, Location and Equipment.
NFPA 88A - 73	Standard for Parking Structures.
NFPA 88B - 73	Standard for Repair Garages.
NFPA 91 - 73	Standard for the Installation of Blower and Exhaust Systems for Dust, Stock, and Vapor Removal, or Conveying.
NFPA 101 - 78	Code for Safety to Life from Fire in Buildings and Structures. (Life Safety Code.)
NFPA 325M - 69	Fire-Hazard Properties of Flammable Liquids, Gases, and Volatile Solids.
NFPA 493 - 75	Standard for Intrinsically Safe Apparatus for Use in Class I Hazardous Locations and Its Associated Apparatus.
NFPA 496 - 74	Standard for Purged and Pressurized Enclosures for Electrical Equipment in Hazardous Locations.
NFPA 497 - 75	Recommended Practice for Classification of

	Class I Hazardous Locations for Electrical Installations in Chemical Plants.
NFPA 505 - 75	Fire Safety Standard for Powered Industrial Trucks Including Type Designations and Areas of Use.
NMAB 353 - 1 - 79	Matrix of Combustion-Relevant Properties and Classification of Gases, Vapors, and Selected Solids.
NMAB 353 - 2 - 79	Test Equipment for Use in Determining Classifications of Combustible Dusts.
NMAB 353 - 3 - 80	Classification of Combustible Dusts in Accordance with the National Electrical Code.